MATHEMATICAL ANALYSIS
A Concise Introduction

MATHEMATICAL ANALYSIS

A Concise Introduction

Jiongmin Yong

University of Central Florida, USA

 **World Scientific**

NEW JERSEY · LONDON · SINGAPORE · BEIJING · SHANGHAI · HONG KONG · TAIPEI · CHENNAI · TOKYO

Published by

World Scientific Publishing Co. Pte. Ltd.

5 Toh Tuck Link, Singapore 596224

USA office: 27 Warren Street, Suite 401-402, Hackensack, NJ 07601

UK office: 57 Shelton Street, Covent Garden, London WC2H 9HE

Library of Congress Cataloging-in-Publication Data

Names: Yong, J. (Jiongmin), 1958–   author.

Title: Mathematical analysis : a concise introduction / Jiongmin Yong, University of Central Florida.

Description: New Jersey : World Scientific, [2021] | Includes bibliographical references and index.

Identifiers: LCCN 2020055934 | ISBN 9789811221637 (hardcover) | ISBN 9789811221644 (ebook)

Subjects: LCSH: Mathematical analysis.

Classification: LCC QA300 .Y53 2021 | DDC 515--dc23

LC record available at https://lccn.loc.gov/2020055934

British Library Cataloguing-in-Publication Data

A catalogue record for this book is available from the British Library.

For any available supplementary material, please visit
https://www.worldscientific.com/worldscibooks/10.1142/11859#t=suppl

Printed in Singapore

In the Memory of My Mother

Preface

Mathematical analysis serves as a common foundation for many research areas of pure and applied mathematics. It is also an important and powerful tool used in many other science fields, including physics, chemistry, biology, engineering, finance, economics, to mention just a few. On the other hand, it is a desire that after having taken a sequence of courses on basic calculus and linear algebra, one could spend a reasonable length of time (ideally, say, one semester), to build an advanced base of analysis sufficient for getting into various research fields other than analysis itself, and/or stepping into more advanced level(s) of analysis courses (such as real analysis, complex analysis, differential equations, functional analysis, stochastic analysis, etc.). To meet such a need, I wrote this concise introduction on mathematical analysis for a course that can be completed within one semester.

From the author's viewpoint, mathematical analysis should present a theory for the functions between the best regular ones (meaning analytic functions which are treated in complex variables/analysis) and the worst regular ones (meaning measurable functions which are treated in real analysis). More precisely, this course discusses functions with various continuity, differentiability, Riemann integrability, and relevant topics of limit, series, and uniform convergence.

For the convergence and continuity of functions, the linear (or algebraic) structure of the underlying space is irrelevant. Therefore, we consider functions and sequences in general metric spaces. As a preparation for that, some basic properties of metric spaces are discussed, such as completeness, compactness, connectedness and separability, etc. For differentiability and Riemann integrability of functions, the linear structure of the underlying space is necessary. Therefore, we discuss the problems of differentiation and integration in the m-dimensional Euclidean space.

Series is an independent topic. Partially, it is an application of the previous chapters if one views it from the limit of partial sums. On the other hand, the study of the series convergence has lots of its own feature. Power series and Fourier series make this part of theory very rich. Also, the theory of series can even touch one of the most challenging open questions in mathematical research: the Riemann hypothesis.

Real number system is traditionally included in the course of mathematical analysis. Due to the time limit of the course, we decided to put it in a self-contained appendix at the end of the book so that the interested readers could still be able to read it separately, and it becomes optional for those readers who are only interested in the main body of mathematical analysis other than the real number system.

Finally, problems listed at the end of sections are complementary to the material covered in the main body of the book, and those with asterisks might be a little difficult. Readers could skip them at the first reading.

Jiongmin Yong
Orlando, USA
July, 2020

Contents

Chapter 1

Metric Spaces and Limits for Sequences

1 Metric Spaces

Let us begin with some notions on *set* which is understood as a collection of distinct objects.

Definition 1.1. (i) Let S be a set. If x is an element of S, we denote $x \in S$. If x is not an element of S, we denote $x \notin S$.

(ii) Let A and B be two sets. If

$$x \in B, \qquad \forall x \in A,$$

we call A a *subset* of B, denoted by $A \subseteq B$.

(iii) Let A and B be two sets. We define the *union, intersection* and *difference* of A and B as follows:

$$A \cup B = \{x \mid x \in A, \text{ or } x \in B\},$$
$$A \cap B = \{x \mid x \in A, \text{ and } x \in B\},$$
$$A \setminus B = \{x \mid x \in A, \text{ and } x \notin B\}.$$

If X is a fixed set and all the considered sets are in X, then for any $S \subseteq X$, the set $S^c = X \setminus S$ is called the *complementary set* of S (in X). For any family A_λ ($\lambda \in \Lambda$) of sets, we can define the union $\bigcup_{\lambda \in \Lambda} A_\lambda$ and the intersection $\bigcap_{\lambda \in \Lambda} A_\lambda$ in the similar way as above.

(iv) If a set S contains no element, it is called an *empty set*, denoted by $S = \varnothing$. Otherwise, it is called a *non-empty set*.

The following collects some basic results concerning the operations of sets, whose proofs are straightforward.

Proposition 1.2. (i) *Let* A, B, C *be subsets of* X. *Then*

$$A \cup B = B \cup A, \qquad A \cap B = B \cap A,$$

and
$$A \cup (B \cap C) = (A \cup B) \cap (A \cup C),$$
$$A \cap (B \cup C) = (A \cap B) \cup (A \cap C),$$
$$A \setminus B = A \cap B^c.$$

(ii) (**de Morgan's Law**) Let $\{A_\lambda \mid \lambda \in \Lambda\}$ be a family of subsets in X. Then

$$\left(\bigcup_{\lambda \in \Lambda} A_\lambda \right)^c = \bigcap_{\lambda \in \Lambda} A_\lambda^c, \qquad \left(\bigcap_{\lambda \in \Lambda} A_\lambda \right)^c = \bigcup_{\lambda \in \Lambda} A_\lambda^c.$$

We now introduce the following definition.

Definition 1.3. Let X be a non-empty set. A map $d : X \times X \to \mathbb{R}$ is called a *metric* on X if the following hold:

(a) (positivity) For any $x, y \in X$, $d(x, y) \geqslant 0$ and $d(x, y) = 0$ if and only if $x = y$.

(b) (symmetry) For any $x, y \in X$, $d(x, y) = d(y, x)$.

(c) (triangle inequality) For any $x, y, z \in X$, it holds:

$$d(x, z) \leqslant d(x, y) + d(y, z). \tag{1.1}$$

In this case, (X, d) is called a *metric space*. The metric $d(\cdot, \cdot)$ is also called a *distance* on X.

We present some examples.

Example 1.4. (i) Let $X = \mathbb{R}$ be the set of all real numbers,[1] and

$$\rho(x, y) = |x - y|, \qquad x, y \in \mathbb{R}.$$

Then a straightforward proof shows that $\rho(\cdot, \cdot)$ is a metric on $\mathbb{R}$, which is called the *standard metric* on $\mathbb{R}$, and $(\mathbb{R}, \rho)$ becomes a metric space.

(ii) Let X be a non-empty set. Define $d_0 : X \times X \to \mathbb{R}$ as follows:

$$d_0(x, y) = \begin{cases} 0, & x = y, \\ 1, & x \neq y. \end{cases}$$

It is easy to see that $d_0(\cdot, \cdot)$ is a metric which is called the *discrete metric* on X. Such a metric can always be defined on any non-empty set X. Hence, for any non-empty set X, one can always define a metric so that it becomes a metric space.

---

[1] See the Appendix for a theory of the real number system.

Example 1.5. Let (X, d) be a metric space and $Y \subseteq X$. Then the *restriction* $d|_{Y \times Y}$ of d on $Y \times Y$ is clearly a metric on Y; $(Y, d|_{Y \times Y})$ is called a *subspace* of (X, d); and $d|_{Y \times Y}$ is called the metric on Y *induced* by d. For convenience, hereafter, we simply denote (Y, d) instead of $(Y, d|_{Y \times Y})$. Under the standard metric, $\mathbb{N} = \{1, 2, 3, \cdots\}$ (the set of all natural numbers), $\mathbb{Z}$ (the set of all integers), $\mathbb{Q}$ (the set of all rational numbers), and interval $[a, b]$ are all subspaces of $\mathbb{R}$.

To have more interesting examples, for any integer $m \geqslant 1$, we define

$$\mathbb{R}^m = \{x \equiv (x^1, x^2, \cdots, x^m) \mid x^k \in \mathbb{R}, \ 1 \leqslant k \leqslant m\}.$$

For any $x = (x^1, x^2, \cdots, x^m), y = (y^1, y^2, \cdots, y^m) \in \mathbb{R}^m$ and $\lambda \in \mathbb{R}$, we introduce

$$x + y = (x^1 + y^1, x^2 + y^2, \cdots, x^m + y^m), \quad \lambda x = (\lambda x^1, \lambda x^2, \cdots, \lambda x^m).$$

These operations are called *addition* and *scalar multiplication*, respectively, for which the following eight *axioms* are satisfied:

$$\begin{cases} x + y = y + x, & \forall x, y \in \mathbb{R}^m, \\ (x + y) + z = x + (y + z), & \forall x, y, z \in \mathbb{R}^m, \\ x + 0 = 0 + x = x, & \forall x \in \mathbb{R}^m, \\ x + (-x) = 0, & \forall x \in \mathbb{R}^m, \\ \lambda(x + y) = \lambda x + \lambda y, & \forall x, y \in \mathbb{R}^m, \ \lambda \in \mathbb{R}, \\ (\lambda + \mu)x = \lambda x + \mu x, & \forall x \in \mathbb{R}^m, \ \lambda, \mu \in \mathbb{R}, \\ 1x = x, & \forall x \in \mathbb{R}^m, \\ \lambda(\mu x) = (\lambda \mu)x, & \forall x \in \mathbb{R}^m, \ \lambda, \mu \in \mathbb{R}. \end{cases}$$

With the addition and scalar multiplication, $\mathbb{R}^m$ is called a *linear space* (or a *vector space*). Each $x \in \mathbb{R}^m$ is called a *vector*. The addition and scalar multiplication are called *algebraic operations* on $\mathbb{R}^m$. Any properties of $\mathbb{R}^m$ involving addition and scalar multiplication are referred to as *algebraic properties* of $\mathbb{R}^m$.

Now, for any $x = (x^1, x^2, \cdots, x^m) \in \mathbb{R}^m$, we define

$$\|x\|_p = \begin{cases} \left(\sum_{k=1}^m |x^k|^p \right)^{\frac{1}{p}}, & p \in [1, \infty), \\ \max_{1 \leqslant k \leqslant m} |x^k|, & p = \infty, \end{cases} \tag{1.2}$$

which is called the ℓ^p-*norm*. We refer to $(\mathbb{R}^m, \|\cdot\|_p)$ as a *normed linear space*. The following proposition is concerned with the ℓ^p-norm.

Proposition 1.6. *Let $p \in [1, \infty]$. Then the above-defined ℓ^p-norm satisfies the following:*

(i) (positivity) $\|x\|_p \geqslant 0$ *for all* $x \in \mathbb{R}^m$, *and* $\|x\|_p = 0$ *if and only if* $x = 0$.

(ii) (positive homogeneity) $\|\lambda x\|_p = |\lambda| \|x\|_p$, *for all* $x \in \mathbb{R}^m$ *and* $\lambda \in \mathbb{R}$.

(iii) (triangle inequality) $\|x + y\|_p \leqslant \|x\|_p + \|y\|_p$, *for all* $x, y \in \mathbb{R}^m$.

From the definition of ℓ^p-norm, we easily see that (i)–(ii) are true. To prove (iii), we first look at the cases $p = 1, \infty$. One has

$$\|x + y\|_1 = \sum_{k=1}^{m} |x^k + y^k| \leqslant \sum_{k=1}^{m} |x^k| + \sum_{k=1}^{m} |y^k| = \|x\|_1 + \|y\|_1,$$

and

$$\|x + y\|_\infty = \max_{1 \leqslant k \leqslant m} |x^k + y^k|$$
$$\leqslant \max_{1 \leqslant k \leqslant m} |x^k| + \max_{1 \leqslant k \leqslant m} |y^k| = \|x\|_\infty + \|y\|_\infty.$$

Now, we look at the case $p = 2$. If $y = 0$, the proof is trivial. Thus, let $x, y \in \mathbb{R}$ with $y \neq 0$. For any $\lambda \in \mathbb{R}$, one has

$$0 \leqslant \|x - \lambda y\|_2^2 = \sum_{k=1}^{m} (x^k - \lambda y^k)^2 = \|x\|_2^2 - 2\lambda \sum_{k=1}^{m} x^k y^k + \lambda^2 \|y\|_2^2.$$

Taking

$$\lambda = \frac{1}{\|y\|_2^2} \sum_{k=1}^{m} x^k y^k,$$

we obtain

$$0 \leqslant \|x\|_2^2 - \frac{2}{\|y\|_2^2} \Big(\sum_{k=1}^{n} x^k y^k \Big)^2 + \frac{1}{\|y\|_2^2} \Big(\sum_{k=1}^{m} x^k y^k \Big)^2,$$

which leads to

$$\Big| \sum_{k=1}^{m} x^k y^k \Big| \leqslant \|x\|_2 \|y\|_2, \qquad \forall x, y \in \mathbb{R}^m. \tag{1.3}$$

This is called the *Cauchy-Schwarz inequality*. Then it follows that

$$\|x+y\|_2^2 = \sum_{k=1}^m (x^k + y^k)^2 = \sum_{k=1}^m \left(|x^k|^2 + 2x^k y^k + |y^k|^2\right)$$

$$= \|x\|_2^2 + 2\sum_{k=1}^m x^k y^k + \|y\|_2^2 \leqslant \|x\|_2^2 + 2\|x\|_2\|y\|_2 + \|y\|_2^2 = \left(\|x\|_2 + \|y\|_2\right)^2.$$

Consequently, (iii) holds for $p = 2$.

For general $p \in (1,2) \cup (2,\infty)$, the situation is a little more subtle. We need the following results which are interesting by themselves.

Proposition 1.7. *Let $p, q \in (1,\infty)$ such that $\frac{1}{p} + \frac{1}{q} = 1$.*

(i) **(Young's Inequality)** *For any $a, b \in (0,\infty)$, the following holds:*

$$ab \leqslant \frac{a^p}{p} + \frac{b^q}{q}. \tag{1.4}$$

(ii) **(Hölder's Inequality)** *For any $x, y \in \mathbb{R}^m$, the following holds:*

$$\left|\sum_{k=1}^m x^k y^k\right| \leqslant \|x\|_p \|y\|_q. \tag{1.5}$$

(iii) **(Minkowski's Inequality)** *For any $x, y \in \mathbb{R}^m$, the following holds:*

$$\|x+y\|_p \leqslant \|x\|_p + \|y\|_p. \tag{1.6}$$

Proof. (i) Let

$$f(r) = \frac{r^p}{p} + \frac{1}{q} - r, \qquad r \in [0,\infty).$$

We try to find the minimum of $f(\cdot)$. To this end,[2] setting

$$f'(r) = r^{p-1} - 1 = 0,$$

we get $r = 1$. Clearly, by the first derivative test, one has

$$f(r) \geqslant f(1) = \frac{1}{p} + \frac{1}{q} - 1 = 0, \quad \forall r \in (0,\infty).$$

This is equivalent to

$$\frac{r^p}{p} + \frac{1}{q} \geqslant r. \tag{1.7}$$

---

[2] We assume that single variable calculus is known to the readers.

Taking $r = ab^{\frac{1}{1-p}}$, one has

$$ab^{\frac{1}{1-p}} = r \leqslant \frac{r^p}{p} + \frac{1}{q} = \frac{a^p b^{\frac{p}{1-p}}}{p} + \frac{1}{q}.$$

Consequently,

$$ab = ab^{\frac{1}{1-p}} b^{1-\frac{1}{1-p}} \leqslant \Big(\frac{a^p b^{\frac{p}{1-p}}}{p} + \frac{1}{q}\Big) b^{\frac{p}{p-1}} = \frac{a^p}{p} + \frac{b^q}{q},$$

which gives (1.4).

(ii) Let

$$A = \|x\|_p = \Big(\sum_{k=1}^m |x^k|^p\Big)^{\frac{1}{p}}, \qquad B = \|y\|_q = \Big(\sum_{k=1}^m |y^k|^q\Big)^{\frac{1}{q}}.$$

If $AB = 0$, then $x = 0$ or $y = 0$ which leads to the conclusion trivially. We now let $AB \neq 0$. Denote

$$a^k = \frac{|x^k|}{A}, \qquad b^k = \frac{|y^k|}{B}, \qquad k = 1, 2, \cdots, m.$$

Then by Young's inequality,

$$a^k b^k \leqslant \frac{(a^k)^p}{p} + \frac{(b^k)^q}{q},$$

which results in

$$\frac{1}{AB}\Big|\sum_{k=1}^m x^k y^k\Big| \leqslant \sum_{k=1}^m a^k b^k \leqslant \frac{1}{pA^p}\sum_{k=1}^m |x^k|^p + \frac{1}{qB^q}\sum_{k=1}^m |y^k|^q$$

$$= \frac{1}{p\|x\|_p^p}\|x\|_p^p + \frac{1}{q\|y\|_q^q}\|y\|_q^q = \frac{1}{p} + \frac{1}{q} = 1.$$

Thus,

$$\Big|\sum_{k=1}^m x^k y^k\Big| \leqslant AB = \|x\|_p \|y\|_q,$$

proving (1.5).

(iii) In the case that $\|x + y\|_p = 0$, the conclusion is trivial. We assume $\|x + y\|_p > 0$ below. By Hölder's inequality,

$$\|x + y\|_p^p = \sum_{k=1}^m |x^k + y^k|^p = \sum_{k=1}^m |x^k + y^k|^{p-1} |x^k + y^k|$$

$$\leqslant \sum_{k=1}^m |x^k + y^k|^{p-1} |x^k| + \sum_{k=1}^m |x^k + y^k|^{p-1} |y^k|$$

$$\leqslant \|x\|_p \Big(\sum_{k=1}^m |x^k + y^k|^{(p-1)q}\Big)^{\frac{1}{q}} + \|y\|_p \Big(\sum_{k=1}^m |x^k + y^k|^{(p-1)q}\Big)^{\frac{1}{q}}$$

$$= \Big(\|x\|_p + \|y\|_p\Big)\Big(\sum_{k=1}^m |x^k + y^k|^p\Big)^{\frac{1}{q}} = \Big(\|x\|_p + \|y\|_p\Big)\|x + y\|_p^{\frac{p}{q}}.$$

Dividing $\|x+y\|_p^{\frac{p}{q}} > 0$ in the above, we get

$$\|x\|_p + \|y\|_p \geqslant \|x+y\|_p^{p-\frac{p}{q}} = \|x+y\|_p.$$

This completes the proof. □

Although in Proposition 1.7, the Hölder's and the Minkowski's inequalities are proved only for the cases $p, q \in (1, \infty)$ with $\frac{1}{p} + \frac{1}{q} = 1$, the cases $p = 1$ (then $q = \infty$) and $p = \infty$ (then $q = 1$) are also true with very simple proofs.

Note that for any $p \in [1, \infty)$, the following estimates hold:

$$\|x\|_\infty = \Big(\max_{1 \leqslant k \leqslant m} |x^k|^p \Big)^{\frac{1}{p}} \leqslant \Big(\sum_{k=1}^m |x^k|^p \Big)^{\frac{1}{p}} = \|x\|_p$$

$$\leqslant \Big(m \max_{1 \leqslant k \leqslant m} |x^k|^p \Big)^{\frac{1}{p}} = m^{\frac{1}{p}} \|x\|_\infty, \qquad \forall x \in \mathbb{R}^m.$$

Consequently, for any $1 \leqslant p, q \leqslant \infty$,

$$m^{-\frac{1}{q}} \|x\|_q \leqslant \|x\|_\infty \leqslant \|x\|_p \leqslant m^{\frac{1}{p}} \|x\|_\infty \leqslant m^{\frac{1}{p}} \|x\|_q, \qquad \forall x \in \mathbb{R}^m. \quad (1.8)$$

This is referred to as the *strong equivalence* among all the ℓ^p-norms (in $\mathbb{R}^m$).[3]

Now, we define

$$\rho_p(x, y) = \|x - y\|_p, \qquad x, y \in \mathbb{R}^m, \quad p \in [1, \infty].$$

It is clear that properties (a) and (b) in the definition of metric (see Definition 1.3) are satisfied by $\rho_p(\cdot, \cdot)$. The triangle inequality follows from Minkowski's inequality. Therefore, $\rho_p(\cdot, \cdot)$ is a metric on $\mathbb{R}^m$, which is called the metric *induced* by the ℓ^p-norm. We call ρ_2 the *Euclidean metric*. From (1.8), we see that

$$m^{-\frac{1}{q}} \rho_q(x, y) \leqslant \rho_p(x, y) \leqslant m^{\frac{1}{p}} \rho_q(x, y), \qquad \forall x, y \in X. \quad (1.9)$$

The above is referred to as the strong equivalence among all the metrics ρ_p induced by the ℓ^p-norms (in $\mathbb{R}^m$).[4]

Proposition 1.8. *Let* (X_k, d_k), $k = 1, 2, \cdots, n$, *be* n *metric spaces. Define*

$$X = X_1 \times X_2 \times \cdots \times X_n \equiv \big\{ (x_1, x_2, \cdots, x_n) \mid x_k \in X_k, \ 1 \leqslant k \leqslant n \big\},$$

---

[3]Note that in (1.8), the dimension m of $\mathbb{R}^m$ is involved. Therefore, such an equivalence is not expected for infinite dimensional spaces.

[4]Similar to the ℓ^p-norm, the strong equivalence of all the metric ρ_p only holds in finite-dimensional normed linear spaces.

and

$$d\big((x_1,\cdots,x_n),(y_1,\cdots,y_n)\big) = \sum_{k=1}^{n} d_k(x_k,y_k). \tag{1.10}$$

Then (X,d) is a metric space, called the product of $(X_1,d_1),\cdots,(X_n,d_n)$.

The proof is obvious. We see that the conclusion of Proposition 1.8 remains true if (1.10) is replaced by

$$d\big((x_1,\cdots,x_n),(y_1,\cdots,y_n)\big) = \Big(\sum_{k=1}^{n} d_k(x_k,y_k)^p\Big)^{\frac{1}{p}}, \qquad 1 \leqslant p < \infty,$$

or

$$d\big((x_1,\cdots,x_n),(y_1,\cdots,y_n)\big) = \sup_{1\leqslant k\leqslant n} d_k(x_k,y_k).$$

Let us present the following additional example.

Example 1.9.[5] For $p \in [1,\infty)$, let

$$\ell^p = \Big\{(x^k) \equiv (x^k)_{k\geqslant 1} \mid x^k \in \mathbb{R}, \ k \geqslant 1; \ \sum_{k=1}^{\infty} |x^k|^p < \infty\Big\},$$

and let

$$\ell^\infty = \Big\{(x^k) \equiv (x^k)_{k\geqslant 1} \mid x^k \in \mathbb{R}, \ k \geqslant 1; \ \sup_{k\geqslant 1}|x^k| < \infty\Big\}.$$

We define

$$\rho_p\big((x^k),(y^k)\big) = \Big(\sum_{k=1}^{\infty} |x^k - y^k|^p\Big)^{\frac{1}{p}}, \qquad \forall (x^k),(y^k) \in \ell^p, \quad p \in [1,\infty),$$

and

$$\rho_\infty\big((x^k),(y^k)\big) = \sup_{k\geqslant 1}|x^k - y^k|, \qquad \forall (x^k),(y^k) \in \ell^\infty.$$

Then $\rho_p(\cdot,\cdot)$ is a metric on ℓ^p ($p \in [1,\infty]$). In fact, again, properties (a) and (b) in the definition of the metric (Definition 1.3) hold for $\rho_p(\cdot,\cdot)$ in ℓ^p. Now for the triangle inequality, we let $N \geqslant 1$ and from the Minkowski's inequality, one has

$$\Big(\sum_{k=1}^{N} |x^k + y^k|^p\Big)^{\frac{1}{p}} \leqslant \Big(\sum_{k=1}^{N} |x^k|^p\Big) + \Big(\sum_{k=1}^{N} |y^k|^p\Big)^{\frac{1}{p}}.$$

Letting $N \to \infty$, one can rigorously show that

$$\|x + y\|_p \leqslant \|x\|_p + \|y\|_p, \qquad \forall x,y \in \ell^p.$$

Hence, $\rho_p(\cdot,\cdot)$ is a metric on ℓ^p.

---

[5]Since some results on series will be involved, those who are not familiar with this could skip this example until they have read Chapter 6.

We point out that unlike $\mathbb{R}^m$, the spaces ℓ^p are different for different $p \in [1, \infty)$, and all ρ_p are not strongly equivalent.

To conclude this section, let us further look at the case $p = 2$. On $\mathbb{R}^m$, we introduce a binary operation, called an *inner product*:

$$\langle x, y \rangle \equiv x^\top y = \sum_{k=1}^{m} x^k y^k, \qquad \forall x, y \in \mathbb{R}^m.$$

With such an inner product, $\mathbb{R}^m$ is called an *inner product space*. Clearly, $(x, y) \mapsto \langle x, y \rangle$ satisfies the following conditions:

$$\begin{cases} \langle x, y \rangle = \langle y, x \rangle, \qquad \forall x, y \in \mathbb{R}^m, \\ \langle \lambda x + \mu y, z \rangle = \lambda \langle x, z \rangle + \mu \langle y, z \rangle, \qquad \forall \lambda, \mu \in \mathbb{R}, \ x, y, z \in \mathbb{R}^m, \\ \langle x, x \rangle \geqslant 0, \qquad \langle x, x \rangle = 0 \iff x = 0. \end{cases} \quad (1.11)$$

Further, one sees that

$$\|x\|_2 = \sqrt{\langle x, x \rangle}, \qquad \forall x \in \mathbb{R}^m.$$

Therefore, we say that $\|\cdot\|_2$ is the norm induced by the inner product $\langle \cdot, \cdot \rangle$. We call $\|\cdot\|_2$ the *Euclidean norm*.

Exercises

1. Prove the de Morgan's Law: Let $\{A_\lambda \mid \lambda \in \Lambda\}$ be a family of sets in $\mathbb{R}$. Then

$$\left(\bigcup_{\lambda \in \Lambda} A_\lambda \right)^c = \bigcap_{\lambda \in \Lambda} A_\lambda^c, \qquad \left(\bigcap_{\lambda \in \Lambda} A_\lambda \right)^c = \bigcup_{\lambda \in \Lambda} A_\lambda^c.$$

2. Let d_1 and d_2 be two metrics on the non-empty set X. Then for any $\lambda_1, \lambda_2 > 0$, $\lambda_1 d_1 + \lambda_2 d_2$ is also a metric on X.

3. Define

$$d(x, y) = \frac{\|x - y\|_p}{1 + \|x - y\|_p}, \qquad \forall x, y \in \mathbb{R}^m,$$

where $\|\cdot\|_p$ is defined by (1.2). Prove or disprove that $d(\cdot, \cdot)$ is a metric on $\mathbb{R}^m$.

4. Let $d(\cdot, \cdot)$ be defined on $\mathbb{R}^m$ as follows:

$$d(x, y) = \left| \{ k \mid x^k \neq y^k, \ k = 1, 2, \cdots, m \} \right|,$$
$$\forall x = (x^1, \cdots, x^m), y = (y^1, \cdots, y^m) \in \mathbb{R}^m,$$

where $|A|$ stands for the number of elements in A. Then $d(\cdot,\cdot)$ is a metric on $\mathbb{R}^m$.

5. Let X be a non-empty set and let $\widehat{\rho} : X \times X \to [0,\infty)$ be a map satisfying the following:

$$\begin{cases} \widehat{\rho}(x,y) \geqslant 0, \quad \forall x,y \in X; \qquad \widehat{\rho}(x,y) = 0 \iff x = y, \\ \widehat{\rho}(x,y) \leqslant \widehat{\rho}(x,z) + \widehat{\rho}(z,y), \qquad \forall x,y,z \in X. \end{cases}$$

Define
$$d(x,y) = \max\big\{\widehat{\rho}(x,y), \widehat{\rho}(y,x)\big\}, \qquad \forall x,y \in X.$$
Then $d(\cdot,\cdot)$ is a metric on X.

6. Let (X,d) be a metric space and $f : X \to X$ be a one-to-one map. Define
$$\bar{d}(x,y) = d\big(f(x), f(y)\big), \qquad x,y \in X.$$
Then $\bar{d}(\cdot,\cdot)$ is a metric on X.

7*. Let ℓ^1 and ℓ^2 be defined in Example 1.9. Then $\ell^1 \subseteq \ell^2$, and $\ell^1 \neq \ell^2$.

2   Sequences and Limits

We introduce the following definition.

Definition 2.1. Let $\{x_n\}_{n\geqslant 1}$ be a *sequence* in a metric space (X,d).

(i) The sequence is said to be *bounded* if there exists an $\widehat{x} \in X$ and a constant $M > 0$ such that
$$d(x_n, \widehat{x}) \leqslant M, \qquad n \geqslant 1.$$

(ii) The sequence is said to be *Cauchy* if for any $\varepsilon > 0$, there exists an $N \geqslant 1$ such that
$$d(x_n, x_m) < \varepsilon, \qquad n, m \geqslant N.$$

(iii) The sequence is said to be *convergent* to $\bar{x} \in X$ with respect to d if for any $\varepsilon > 0$, there exists an $N \geqslant 1$ such that
$$d(x_n, \bar{x}) < \varepsilon, \qquad n \geqslant N.$$

In this case, $\bar{x}$ is called a *limit* of the sequence $\{x_n\}_{n\geqslant 1}$. The above is also denoted by the following:
$$\lim_{n\to\infty} d(x_n, \bar{x}) = 0.$$
When the metric d is clear from the context, we simply denote the above by
$$\lim_{n\to\infty} x_n = \bar{x}.$$
The sequence is said to be *divergent* if it is not convergent.

From the above definition, we see that a sequence $\{x_n\}_{n \geqslant 1}$ is bounded, Cauchy, convergent, or divergent, if and only if so is the sequence $\{x_n\}_{n \geqslant k}$ for any fixed $k \geqslant 1$. In other words, the above-mentioned properties remain if we drop any first finitely many terms of the sequence. Likewise, by adding finitely many terms at the beginning of the sequence, the above-mentioned properties will not be changed. Hence, from now on, we will simply denote the sequence by $\{x_n\}$ instead of $\{x_n\}_{n \geqslant 1}$, unless the initial term needs to be emphasized.

The following result gives some basic properties of convergent sequences.

Proposition 2.2. *Let (X, d) be a metric space and $\{x_n\} \subset X$ be a sequence.*

(i) *If $\{x_n\}$ is convergent to $x \in X$ and to $x' \in X$, then $x = x'$.*

(ii) *If $\{x_n\}$ is convergent, then it is Cauchy.*

(iii) *If $\{x_n\}$ is Cauchy, then it is bounded.*

Proof. (i) Suppose $\{x_n\}$ has two different limits $x, x' \in X$. Then for $\varepsilon = \frac{d(x, x')}{2} > 0$, there exists an $N \geqslant 1$ such that

$$d(x_n, x) + d(x_n, x') < \varepsilon, \qquad \forall n \geqslant N.$$

Consequently, for any $n \geqslant N$,

$$0 < d(x, x') \leqslant d(x, x_n) + d(x_n, x') < \varepsilon = \frac{d(x, x')}{2},$$

a contradiction.

(ii) Let $\{x_n\}$ converge to $\bar{x}$. Then for any $\varepsilon > 0$, there exists an $N \geqslant 1$ such that

$$d(x_n, \bar{x}) < \frac{\varepsilon}{2}, \qquad \forall n \geqslant N.$$

Consequently, for any $m, n \geqslant N$, we have

$$d(x_n, x_m) \leqslant d(x_n, \bar{x}) + d(x_m, \bar{x}) < \varepsilon.$$

This shows that $\{x_n\}$ is Cauchy.

(iii) Let $\{x_n\}$ be Cauchy. Then for $\varepsilon = 1 > 0$, there exists an $N \geqslant 1$ such that

$$d(x_n, x_N) < \varepsilon = 1, \qquad \forall n \geqslant N.$$

Now letting

$$M = 1 + \sum_{k=1}^{N-1} d(x_k, x_N),$$

one has

$$d(x_n, x_N) \leqslant M, \qquad \forall n \geqslant 1.$$

This completes the proof.   □

The following result is a little surprising, but the proof is obvious, which is left to the readers.

Proposition 2.3. *Let (X, d_0) be a discrete metric space and $\{x_n\} \subset X$ be a sequence. Then $\{x_n\}$ is convergent to $x \in X$ if and only if there exists an $N \geqslant 1$ such that*

$$x_n = x, \qquad \forall n \geqslant N.$$

We point out that by changing the metric, the convergence of sequences may be changed. For example, the sequence $\{\frac{1}{n}\}_{n \geqslant 1}$ converges to 0 under the standard metric of $\mathbb{R}$. However, this sequence is not convergent under the discrete metric. Further, it is even possible that when the metric is changed, the limit of a convergent sequence might be changed. Here is an example.

Example 2.4. Let

$$f(x) = \begin{cases} 1, & x = 0, \\ x, & x \in (0, 1), \\ 0, & x = 1. \end{cases}$$

Let $X = [0, 1]$ and $\rho(x, y) = |x - y|$ be the standard metric. Define

$$\bar{d}(x, y) = \rho(f(x), f(y)) = |f(x) - f(y)|, \qquad \forall x, y \in X.$$

Clearly,

$$\bar{d}(x, y) = |f(x) - f(y)| = |f(y) - f(x)| = \bar{d}(x, y) \geqslant 0, \quad \forall x, y \in [0, 1],$$

and since $f(\cdot)$ is one-to-one, we have

$$\bar{d}(x, y) = |f(x) - f(y)| = 0 \quad \Longleftrightarrow \quad x = y.$$

Finally,

$$\bar{d}(x, y) = |f(x) - f(y)| \leqslant |f(x) - f(z)| + |f(z) - f(y)|$$
$$= \bar{d}(x, z) + \bar{d}(z, y), \qquad \forall x, y, z \in [0, 1].$$

Hence, $\bar{d}(\cdot,\cdot)$ is a metric on X. Now, for sequence $x_n = \frac{1}{n}$, we have

$$\rho\left(\frac{1}{n},0\right) = \frac{1}{n} \to 0,$$

and

$$\bar{d}\left(\frac{1}{n},1\right) = \left|\frac{1}{n} - f(1)\right| = \frac{1}{n} \to 0.$$

Thus the sequence is convergent under ρ and $\bar{d}$, but the limits are different!

Relevant to the above example, we introduce the following definition.

Definition 2.5. Let $d,\bar{d}$ be two metrics on X.

(i) d is said to be *stronger* than $\bar{d}$ if for any sequence $\{x_n\} \subseteq X$, and some $\bar{x} \in X$,

$$\lim_{n\to\infty} d(x_n,\bar{x}) = 0 \quad\Rightarrow\quad \lim_{n\to\infty} \bar{d}(x_n,\bar{x}) = 0.$$

In this case, $\bar{d}$ is said to be *weaker* than d.

(ii) d and $\bar{d}$ are said to be *equivalent* if for any sequence $\{x_n\} \subseteq X$, and some $\bar{x} \in X$,

$$\lim_{n\to\infty} d(x_n,\bar{x}) = 0 \quad\Longleftrightarrow\quad \lim_{n\to\infty} \bar{d}(x_n,\bar{x}) = 0.$$

It is clear that if two metrics d and $\bar{d}$ are strongly equivalent (see (1.9)), then they are equivalent. Hence, in $\mathbb{R}^m$, convergence of sequences under any ρ_p ($p \in [1,\infty]$) are all equivalent. We point out that equivalence of metrics do not have to be strongly equivalent. (Why?)

Proposition 2.6. *Let $\{x_n\}$ be a sequence in $\mathbb{R}^m$ with $x_n = (x_n^1, x_n^2, \cdots, x_n^m)$, and $x = (x^1, x^2, \cdots, x^m) \in \mathbb{R}^m$. Then the following are equivalent:*

(i) *$\{x_n\}$ converges to x with respect to ρ_p, for some $p \in [1,\infty]$;*

(ii) *$\{x_n\}$ converges to x with respect to ρ_p, for every $p \in [1,\infty]$;*

(iii) *For each $1 \leqslant k \leqslant m$, the sequence $\{x_n^k\}$ is convergent to x^k.*

The proof is left to the readers.

Definition 2.7. Let (X,d) be a metric space, and $\{x_n\}_{n\geqslant 1}$ be a sequence in X.

(i) Let $\sigma : \mathbb{N} \to \mathbb{N}$ be a strictly increasing map. Then $\{x_{\sigma(n)}\}_{n\geqslant 1}$ is called a *subsequence* of $\{x_n\}_{n\geqslant 1}$. Sometimes, a subsequence is denoted by $\{x_{n_k}\}_{k\geqslant 1}$.

(ii) A point $\bar{x} \in X$ is called a *limit point* of the sequence $\{x_n\}_{n \geqslant 1}$ if there exists a subsequence $\{x_{\sigma(n)}\}_{n \geqslant 1}$ such that

$$\lim_{n \to \infty} d(x_{\sigma(n)}, \bar{x}) = 0.$$

Proposition 2.8. *Let (X, d) be a metric space, and $\{x_n\} \subset X$ be a sequence. If $\{x_n\}$ is convergent to $\bar{x}$, then any subsequence $\{x_{\sigma(n)}\}$ of $\{x_n\}$ is convergent to $\bar{x}$ as well. Conversely, if every subsequence $\{x_{\sigma(n)}\}$ of $\{x_n\}$ has a convergent sub-subsequence with all the limits being $\bar{x}$, then the sequence $\{x_n\}$ is convergent to $\bar{x}$.*

Proof. Suppose $\{x_n\}$ is convergent to $\bar{x}$. Then for any $\varepsilon > 0$, there exists an $N \geqslant 1$ such that

$$d(x_n, \bar{x}) < \varepsilon, \qquad \forall n \geqslant N.$$

Now, let $\{x_{\sigma(n)}\}$ be any subsequence of $\{x_n\}$ (with $\sigma : \mathbb{N} \to \mathbb{N}$ being a strictly increasing map). Noting $\sigma(n) \geqslant n$, we have

$$d(x_{\sigma(n)}, \bar{x}) < \varepsilon, \qquad \forall n \geqslant N.$$

Thus, $\{x_{\sigma(n)}\}$ is convergent to $\bar{x}$.

Conversely, we prove by contradiction. Suppose $\{x_n\}$ is not convergent. Then there exists an $\varepsilon > 0$ and two subsequences $\{x_{\sigma_1(n)}\}$ and $\{x_{\sigma_2(n)}\}$ such that

$$d(x_{\sigma_1(n)}, x_{\sigma_2(n)}) \geqslant \varepsilon, \qquad n \geqslant 1.$$

Consequently, the limits of the convergent subsequences of $\{x_{\sigma_1(n)}\}$ and $\{x_{\sigma_2(n)}\}$ will be different, a contradiction. $\qquad \square$

For the case $X = \mathbb{R}$, we have some additional notions and results concerning the limits. For convenience, we let $\bar{c}$, c, and ℓ^∞ be the sets of all convergent, Cauchy, bounded sequences in $\mathbb{R}$, respectively, and let ℓ^0 be the set of all general sequences in $\mathbb{R}$. Then from Proposition 2.2, one has

$$\bar{c} \subseteq c \subseteq \ell^\infty \subseteq \ell^0.$$

Let us now state the following result whose proof follows from the definition of the limit easily.

Proposition 2.9. *Let $\{x_n\}, \{y_n\} \in \bar{c}$. Then*

$$\lim_{n \to \infty} (x_n \pm y_n) = \lim_{n \to \infty} x_n \pm \lim_{n \to \infty} y_n,$$

$$\lim_{n \to \infty} (x_n y_n) = (\lim_{n \to \infty} x_n)(\lim_{n \to \infty} y_n),$$

$$\lim_{n \to \infty} \frac{x_n}{y_n} = \frac{\lim_{n \to \infty} x_n}{\lim_{n \to \infty} y_n}, \quad \text{if } \lim_{n \to \infty} y_n \neq 0.$$

The following gives a comparison theorem of sequences and their limits.

Proposition 2.10. *Let* $\{x_n\}, \{y_n\} \in \bar{c}$ *and let*

$$x_n \leqslant y_n, \qquad \forall n \geqslant N_0,$$

for some $N_0 \geqslant 1$. *Then*

$$\lim_{n \to \infty} x_n \leqslant \lim_{n \to \infty} y_n. \tag{2.1}$$

Proof. Let

$$\bar{x} = \lim_{n \to \infty} x_n, \qquad \bar{y} = \lim_{n \to \infty} y_n.$$

Suppose $\bar{x} > \bar{y}$ instead, then for $\varepsilon = \frac{\bar{x} - \bar{y}}{2} > 0$, there exists an $N \geqslant 1$, such that

$$|x_n - \bar{x}| + |y_n - \bar{y}| < \frac{\varepsilon}{2}, \qquad n \geqslant N.$$

This leads to (noting $x_n - y_n \leqslant 0$)

$$0 < 2\varepsilon < \bar{x} - \bar{y} = \bar{x} - x_n + x_n - y_n + y_n - \bar{y}$$
$$\leqslant |\bar{x} - x_n| + |y_n - \bar{y}| < \varepsilon, \qquad n \geqslant N,$$

a contradiction. $\qquad \qquad \square$

The above result has a very useful corollary whose proof is obvious.

Corollary 2.11. (Squeeze Theorem) *Suppose* $\{x_n\}, \{y_n\}, \{z_n\} \in \ell^0$ *satisfy*

$$x_n \leqslant y_n \leqslant z_n, \qquad \forall n \geqslant N_0,$$

for some $N_0 \geqslant 1$. *Further, suppose*

$$\lim_{n \to \infty} x_n = \lim_{n \to \infty} z_n = \bar{x}.$$

Then

$$\lim_{n \to \infty} y_n = \bar{x}.$$

Next, we consider sequence with *infinite limit*.

Definition 2.12. Let $\{x_n\} \in \ell^0$. If for any $M > 0$, there exists an $N \geqslant 1$ such that

$$x_n > M, \qquad \forall n \geqslant N,$$

then we say that x_n approaches to ∞ as n goes to infinity, denoted by $x_n \to \infty$, or

$$\lim_{n \to \infty} x_n = \infty.$$

Likewise, if for any $M > 0$, there exists an $N \geqslant 1$ such that

$$x_n < -M, \qquad \forall n \geqslant N,$$

then we say that x_n approaches to $-\infty$ as n goes to infinity, denoted by $x_n \to -\infty$, or

$$\lim_{n \to \infty} x_n = -\infty.$$

The following comparison of sequences is obvious. We leave the proof to the readers.

Proposition 2.13. *Let* $\{x_n\}, \{y_n\} \in \ell^0$ *satisfying*

$$x_n \leqslant y_n, \qquad \forall n \geqslant N_0,$$

for some $N_0 \geqslant 1$. *Then*

$$x_n \to \infty \qquad \Rightarrow \qquad y_n \to \infty,$$
$$y_n \to -\infty \qquad \Rightarrow \qquad x_n \to -\infty.$$

Monotone sequences are of special importance and they will play useful roles below.

Proposition 2.14. *Let* $\{x_n\} \in \ell^0$.

(i) *Suppose there exists an* $M \in \mathbb{R}$ *such that*

$$x_n \leqslant x_{n+1} \leqslant M, \qquad \forall n \geqslant 1.$$

Then $\{x_n\} \in \bar{c}$, *and*

$$\lim_{n \to \infty} x_n = \sup_{n \geqslant 1} x_n \leqslant M,$$

where $\sup_{n \geqslant 1} x_n$ *is called the* supremum *(or the* least upper bound*) of the sequence* $\{x_n\}_{n \geqslant 1}$.

(ii) *Suppose there exists an* $M \in \mathbb{R}$ *such that*

$$x_n \geqslant x_{n+1} \geqslant M, \qquad \forall n \geqslant 1.$$

Then $\{x_n\} \in \bar{c}$, *and*

$$\lim_{n \to \infty} x_n = \inf_{n \geqslant 1} x_n \geqslant M,$$

where $\inf_{n \geqslant 1} x_n$ *is called the* infimum *(or the* greatest lower bound*) of the sequence* $\{x_n\}_{n \geqslant 1}$.[6]

---

[6]The existence of supremum and infimum relies on the theory of the real number system, which will be presented in the Appendix.

Proof. (i) First of all, since $\{x_n\}$ is bounded above, by a result from the Appendix,

$$b \equiv \sup_{n \geq 1} x_n \in \mathbb{R}$$

exists. Now, for any $\varepsilon > 0$, there exists an $N \geq 1$ such that

$$x_N > b - \varepsilon.$$

By the monotonicity of $\{x_n\}$, we have

$$b - \varepsilon < x_N \leq x_n \leq b, \qquad \forall n \geq N.$$

This implies

$$|x_n - b| < \varepsilon, \qquad \forall n \geq N.$$

Hence, we have the convergence of x_n to b.

(ii) The proof is similar and is left to the readers. $\qquad\qquad\square$

We know that $\bar{c} \subseteq c \subset \ell^\infty$, and $c \neq \ell^\infty$. A natural question is what can we say a little more about sequences in $\ell^\infty \setminus c$? We now investigate that.

Let $\{x_n\}_{n \geq 1} \in \ell^\infty$. Then for any $k \geq 1$, $\{x_n\}_{n \geq k} \in \ell^\infty$, and the following is well-defined:

$$y_k = \sup_{n \geq k} x_n.$$

It is not hard to show that $\{y_k\}_{k \geq 1}$ is bounded and non-increasing:

$$y_k \geq y_{k+1}, \qquad \forall k \geq 1.$$

Thus, by Proposition 2.14, the following limit exists:

$$\lim_{k \to \infty} y_k = \inf_{k \geq 1} y_k = \inf_{k \geq 1} \sup_{n \geq k} x_n = \lim_{k \to \infty} \sup_{n \geq k} x_n \equiv \overline{\lim_{n \to \infty}} \, x_n.$$

This is called the *limit superior* (limsup, for short) of $\{x_n\}_{n \geq 1}$.

Likewise, we may define

$$z_k = \inf_{n \geq k} x_n, \qquad k \geq 1,$$

which is a non-decreasing bounded sequence. Hence, the following limit exists, by Proposition 2.14 again:

$$\lim_{k \to \infty} z_k = \sup_{k \geq 1} z_k = \sup_{k \geq 1} \inf_{n \geq k} x_n = \lim_{k \to \infty} \inf_{n \geq k} x_n \equiv \underline{\lim}_{n \to \infty} \, x_n.$$

This is called the *limit inferior* (liminf, for short) of $\{x_n\}_{n \geq 1}$.

The above shows that for any $\{x_n\} \in \ell^\infty$, $\varlimsup\limits_{n\to\infty} x_n$ and $\varliminf\limits_{n\to\infty} x_n$ always exist, thanks to the theory of the real number system, which guarantees the existence of supremum and infimum for bounded sequences (of real numbers). Further, if $\{x_n\}_{n\geqslant 1}$ is unbounded, $\varlimsup\limits_{n\to\infty} x_n$ and $\varliminf\limits_{n\to\infty} x_n$ can still be defined if they are allowed to be $\pm\infty$.

The following gives some relations among limsup, liminf, and limit.

Proposition 2.15. *For any sequence $\{x_n\} \in \ell^\infty$, the following hold:*

$$\varliminf_{n\to\infty} x_n \leqslant \varlimsup_{n\to\infty} x_n, \qquad \varlimsup_{n\to\infty} (-x_n) = -\varliminf_{n\to\infty} x_n. \tag{2.2}$$

Further, $\lim\limits_{n\to\infty} x_n$ *exists if and only if*

$$\varliminf_{n\to\infty} x_n = \varlimsup_{n\to\infty} x_n. \tag{2.3}$$

In this case,

$$\lim_{n\to\infty} x_n = \varliminf_{n\to\infty} x_n = \varlimsup_{n\to\infty} x_n.$$

In particular,

$$\lim_{n\to\infty} x_n = 0 \qquad \Longleftrightarrow \qquad \varlimsup_{n\to\infty} |x_n| = 0. \tag{2.4}$$

Proof. By definition, we have

$$z_k = \inf_{n\geqslant k} x_n \leqslant \sup_{n\geqslant k} x_n = y_k, \qquad \forall k \geqslant 1.$$

Hence,

$$\varliminf_{n\to\infty} x_n = \lim_{k\to\infty} z_k \leqslant \lim_{k\to\infty} y_k = \varlimsup_{n\to\infty} x_n.$$

This proves the first in (2.2). The second in (2.2) can be proved similarly.

Next, suppose $\lim\limits_{n\to\infty} x_n = \bar{x}$. Then for any $\varepsilon > 0$, there exists an $N \geqslant 1$ such that

$$\bar{x} - \varepsilon < x_n < \bar{x} + \varepsilon, \qquad \forall n \geqslant N.$$

Hence,

$$\bar{x} - \varepsilon \leqslant z_k \leqslant y_k \leqslant \bar{x} + \varepsilon, \qquad \forall k \geqslant N.$$

Consequently,

$$\bar{x} - \varepsilon \leqslant \varliminf_{n\to\infty} x_n \leqslant \varlimsup_{n\to\infty} x_n \leqslant \bar{x} + \varepsilon.$$

Since $\varepsilon > 0$ is arbitrary, we obtain

$$\varliminf_{n\to\infty} x_n = \varlimsup_{n\to\infty} x_n = \bar{x} = \lim_{n\to\infty} x_n.$$

Conversely, if

$$\varliminf_{n\to\infty} x_n = \varlimsup_{n\to\infty} x_n = \bar{x},$$

then for any $\varepsilon > 0$, there exists an $N \geqslant 1$ such that

$$|y_k - \bar{x}| < \varepsilon, \quad |z_k - \bar{x}| < \varepsilon, \qquad \forall k \geqslant N,$$

with

$$\bar{x} - \varepsilon < z_k = \inf_{\ell \geqslant k} x_\ell \leqslant x_n \leqslant \sup_{\ell \geqslant k} x_\ell = y_k < \bar{x} + \varepsilon, \quad n \geqslant k \geqslant N.$$

This leads to

$$|x_n - \bar{x}| < \varepsilon, \qquad \forall n \geqslant N.$$

Hence,

$$\lim_{n\to\infty} x_n = \bar{x}.$$

We leave the proof of the rest of the conclusions to the readers. $\quad\square$

Proposition 2.16. *Let* $\{x_n\}, \{y_n\} \in \ell^\infty$.

(i) *The following hold:*

$$\varlimsup_{n\to\infty} (x_n + y_n) \leqslant \varlimsup_{n\to\infty} x_n + \varlimsup_{n\to\infty} y_n,$$

$$\varliminf_{n\to\infty} (x_n + y_n) \geqslant \varliminf_{n\to\infty} x_n + \varliminf_{n\to\infty} y_n.$$

(ii) *Let*

$$x_n \leqslant y_n, \qquad \forall n \geqslant 1.$$

Then

$$\varlimsup_{n\to\infty} x_n \leqslant \varlimsup_{n\to\infty} y_n, \qquad \varliminf_{n\to\infty} x_n \leqslant \varliminf_{n\to\infty} y_n.$$

The proof is left to the readers. Let us now recall the definition of limit point (see Definition 2.7 (ii)).

Proposition 2.17. *Let* $\{x_n\} \in \ell^\infty$. *Then both* $\varlimsup_{n\to\infty} x_n$ *and* $\varliminf_{n\to\infty} x_n$ *are limit points of* $\{x_n\}$. *Further, if* $\bar{x}$ *is a limit point of* $\{x_n\}_{n\geqslant 1}$, *then*

$$\varliminf_{n\to\infty} x_n \leqslant \bar{x} \leqslant \varlimsup_{n\to\infty} x_n.$$

The proof is left to the readers.

Exercises

1. Find an example showing that a bounded sequence does not have to be Cauchy.

2. Let $\{x_n\}_{n \geqslant 1}$ be Cauchy. Then either for any $N \geqslant 1$, there exists an $\bar{n} \geqslant N$ such that

$$x_n \leqslant x_{\bar{n}}, \qquad \forall n \geqslant N,$$

or there exists an $\bar{n} \geqslant N$ such that

$$x_n \geqslant x_{\bar{n}}, \qquad \forall n \geqslant N.$$

Is it possible that for any $N \geqslant 1$ there always exist $n_1, n_2 \geqslant N$ such that

$$x_{n_1} \leqslant x_n \leqslant x_{n_2}, \qquad \forall n \geqslant N?$$

3. If $\{x_n\}, \{y_n\} \in c$, then $\{\max(x_n, y_n)\} \in c$.

4. Let c_0 be the set of all real sequences that converge to 0. Then c_0 is a linear space, i.e., for any $\{x_n\}, \{y_n\} \in c_0$ and any $\alpha, \beta \in \mathbb{R}$, $\{\alpha x_n + \beta y_n\} \in c_0$. Is the conclusion true for c (the set of all Cauchy sequences)? Why?

5. Let $\{x_n\}, \{y_n\} \in \ell^\infty$ satisfy

$$x_n < y_n, \qquad n \geqslant 1.$$

Then

$$\varlimsup_{n \to \infty} x_n \leqslant \varlimsup_{n \to \infty} y_n, \qquad \varliminf_{n \to \infty} x_n \leqslant \varliminf_{n \to \infty} y_n.$$

For each of the above, find examples that the equalities hold.

6. Let $\{x_n\}, \{y_n\} \in \ell^\infty$. Then

$$\varlimsup_{n \to \infty} (x_n + y_n) \leqslant \varlimsup_{n \to \infty} x_n + \varlimsup_{n \to \infty} y_n,$$
$$\varliminf_{n \to \infty} (x_n + y_n) \geqslant \varliminf_{n \to \infty} x_n + \varliminf_{n \to \infty} y_n.$$

Further, for each of the above, find an example that the strict inequalities hold.

7. Let $\{x_n\} \in c$ and $\{y_n\} \in \ell^\infty$. Then

$$\varlimsup_{n \to \infty} (x_n + y_n) = \lim_{n \to \infty} x_n + \varlimsup_{n \to \infty} y_n.$$

8. Let $\{x_n\}_{n \geqslant 1}, \{y_n\}_{n \geqslant 1}, \{z_n\}_{n \geqslant 1} \in \ell^\infty$ such that

$$x_n \leqslant y_n \leqslant z_n, \qquad \forall n \geqslant 1.$$

Further, let the following hold:

$$\varlimsup_{n \to \infty} x_n = \varlimsup_{n \to \infty} z_n = \overline{L}, \quad \varliminf_{n \to \infty} x_n = \varliminf_{n \to \infty} z_n = \underline{L}.$$

Then

$$\varlimsup_{n \to \infty} y_n = \overline{L}, \quad \varliminf_{n \to \infty} y_n = \underline{L}.$$

From the above, prove the Squeeze Theorem (Corollary 2.11).

9. Let $0 < a < b$, $x_1 = a > 0$, $y_1 = b > 0$, and

$$x_{n+1} = \sqrt{x_n y_n}, \qquad y_{n+1} = \frac{x_n + y_n}{2}, \qquad n \geqslant 2.$$

Then $\lim\limits_{n \to \infty} x_n$ and $\lim\limits_{n \to \infty} y_n$ exist and they are equal.

10. Let $\lim\limits_{n \to \infty} x_n = a$. Then $\lim\limits_{n \to \infty} \dfrac{x_1 + x_2 + \cdots + x_n}{n} = a$.

11. Let $x_0 = 1$, $x_{n+1} = 1 + \frac{x_n}{1 + x_n}$, $n \geqslant 0$. Show that $\lim\limits_{n \to \infty} x_n$ exists and further, find the limit.

12. Suppose $\{x_n\} \in \ell^\infty \setminus c$. Show that there are two convergent subsequences of $\{x_n\}$ having different limits.

13. Let $\{x_n\} \in c$ such that for some subsequence $\{x_{n_k}\}_{k \geqslant 1}$, it holds $\lim\limits_{k \to \infty} x_{n_k} = \bar{x}$. Then $\lim\limits_{n \to \infty} x_n = \bar{x}$.

14. Let (X, d_0) be a discrete metric space. Let $x_n \in X$ be a sequence converging to $\bar{x}$ under d_0. Then there exists an $N \geqslant 1$ such that

$$x_n = \bar{x}, \qquad n \geqslant N.$$

3 Sets in Metric Spaces

In this section, we look at some sets in metric spaces, and their properties. Let us first introduce the following definition.

Definition 3.1. Let (X, d) be a metric space.

(i) Let $x_0 \in X$ and $r > 0$. Define

$$B(x_0, r) \equiv B_{(X,d)}(x_0, r) = \{x \in X \mid d(x, x_0) < r\},$$

which is called the *open ball* centered at x_0 with radius r.

(ii) A point $x_0 \in E \subseteq X$ is called an *interior point* of E if there exists a $\delta > 0$ such that

$$B(x_0, \delta) \subseteq E. \tag{3.1}$$

The set of all interior points of E is denoted by E°, which is called the *interior* of E.

(iii) A point $x_0 \in X$ is called a *boundary point* of E if for any $\delta > 0$,

$$B(x_0, \delta) \cap E \neq \varnothing, \qquad B(x_0, \delta) \cap E^c \neq \varnothing.$$

The set of all boundary points of E is denoted by ∂E, which is called the *boundary* of E.

(iv) A point $x_0 \in X$ is called an *adherent point* of E if for any $\delta > 0$,

$$E \cap B(x_0, \delta) \neq \varnothing.$$

The set of all adherent points of E is denoted by $\bar{E}$, which is called the *closure* of E.

(v) A point $x_0 \in X$ is called a *limit point* of E if it is an adherent point of $E \setminus \{x_0\}$.

(vi) A point $x_0 \in E$ is called an *isolated point* of E if there exists a $\delta > 0$ such that

$$\Big(E \setminus \{x_0\} \Big) \cap B(x_0, \delta) = \varnothing.$$

(vii) A set E is said to be *open* if

$$E = E^\circ,$$

i.e., for any $x_0 \in E$, there exists a $\delta > 0$ such that (3.1) holds, which is also equivalent to

$$E \cap \partial E = \varnothing.$$

(viii) A set E is said to be *closed* if

$$\bar{E} = E, \tag{3.2}$$

which is equivalent to

$$\partial E \subseteq E. \tag{3.3}$$

(ix) A subset E of F is said to be *dense* in F if $\bar{E} \supseteq F$.

(x) A set E is said to be *disconnected* if there are two non-empty sets $G_1, G_2 \subseteq X$ such that

$$E = G_1 \cup G_2, \quad \bar{G}_1 \cap G_2 = \varnothing, \quad G_1 \cap \bar{G}_2 = \varnothing. \tag{3.4}$$

If E is not disconnected, we say that E is *connected*. In the case $E = X$, we call that X is a *disconnected metric space* and *connected metric space*, respectively, if $E = X$ is disconnected and connected, respectively.

For the connectedness above, we look at the following simple situation: The set $S = (-1,0) \cup (0,1)$ is disconnected. In fact, by letting $G_1 = (-1,0)$ and $G_2 = (0,1)$, we see that the last two relations in (3.4) hold. On the other hand, let us consider $\widetilde{S} = (-1,1)$ which should be not disconnected. By letting $G_1 = (-1,0]$ and $G_2 = (0,1)$, although $\widetilde{S} = G_1 \cup G_2$ and $G_1 \cap G_2 = \varnothing$, but $G_1 \cap \bar{G}_2 = \{0\} \neq \varnothing$, i.e., the last relation in (3.4) fails. The point that we want to make here is that $G_1 \cap G_2 = \varnothing$ is not enough for disconnectedness.

Example 3.2. (i) Any finite set A is closed, and all of its points are isolated points.

(ii) The sets $\varnothing$, $\mathbb{N}$, $\mathbb{Z}$, and $\mathbb{R}$ are closed. All points in $\mathbb{N}$ and $\mathbb{Z}$ are isolated, and all points in $\mathbb{R}$ are limit points.

(iii) The set $\mathbb{Q}$ is not closed and its closure $\bar{\mathbb{Q}} = \mathbb{R}$. Therefore, $\mathbb{Q}$ is dense in $\mathbb{R}$. All points in $\mathbb{Q}$ are limit points of $\mathbb{Q}$. There are no isolated points in $\mathbb{Q}$.

Proposition 3.3. *Let (X,d) be a metric space.*

(i) *Let $S_1 \subseteq S_2 \subseteq X$. Then*

$$\bar{S}_1 \subseteq \bar{S}_2, \qquad S_1^\circ \subseteq S_2^\circ. \tag{3.5}$$

(ii) *Let $S_1, S_2 \subseteq X$ be two sets. Then*

$$\overline{(S_1 \cap S_2)} \subseteq \bar{S}_1 \cap \bar{S}_2, \qquad \overline{(S_1 \cup S_2)} = \bar{S}_1 \cup \bar{S}_2, \tag{3.6}$$

and

$$(S_1 \cap S_2)^\circ = S_1^\circ \cap S_2^\circ, \qquad (S_1 \cup S_2)^\circ \supseteq S_1^\circ \cup S_2^\circ. \tag{3.7}$$

Proof. The proof of (i) is clear. We now prove (ii). First of all,

$$S_1 \cap S_2 \subset S_1, S_2 \subseteq S_1 \cup S_2. \tag{3.8}$$

Thus, by (i), we have

$$\overline{(S_1 \cap S_2)} \subseteq \bar{S}_1 \cap \bar{S}_2, \qquad \bar{S}_1 \cup \bar{S}_2 \subseteq \overline{(S_1 \cup S_2)}.$$

Now, we prove the equality in (3.6). For any $x \in \overline{(S_1 \cup S_2)}$, there exists a sequence $\{x_n\} \subseteq S_1 \cup S_2$ converging to x. Clearly, there must be a subsequence of $\{x_n\}$ either in S_1 or in S_2. (Why?) Hence, x is either in $\bar{S}_1$ or in $\bar{S}_2$. Therefore,

$$\overline{(S_1 \cup S_2)} = \bar{S}_1 \cup \bar{S}_2.$$

Next, from (3.8), by (i) again, we have

$$(S_1 \cap S_2)^\circ \subseteq S_1^\circ \cap S_2^\circ, \qquad S_1^\circ \cup S_2^\circ \subseteq (S_1 \cup S_2)^\circ.$$

We now prove the equality in (3.7). If there is an $x \in S_1^\circ \cap S_2^\circ$ which is an open set, then there exists a $\delta > 0$ such that

$$B(x, \delta) \subseteq S_1^\circ \cap S_2^\circ \subseteq S_1 \cap S_2.$$

This leads to $x \in (S_1 \cap S_2)^\circ$. Hence,

$$(S_1 \cap S_2)^\circ = S_1^\circ \cap S_2^\circ.$$

This completes the proof of (ii). $\qquad\qquad\qquad\qquad\qquad\qquad\qquad$ □

Note that if we take

$$S_1 = (0, 1), \quad S_2 = (1, 2),$$

then

$$\overline{(S_1 \cap S_2)} = \varnothing \neq \{1\} = \bar{S}_1 \cap \bar{S}_2.$$

Also, if we take

$$S_1 = (0, 1], \qquad S_2 = [1, 2),$$

then

$$S_1^\circ \cup S_2^\circ = (0, 1) \cup (1, 2) \neq (0, 2) = (S_1 \cup S_2)^\circ.$$

Hence, the two inclusions in (3.6) and (3.7) may be strict, in general.

Proposition 3.4. *Let (X, d) be a metric space and $E \subseteq X$. Then*

$$\bar{E} = E \cup \partial E = E^\circ \cup \partial E. \tag{3.9}$$

Proof. First of all, by the definition of adherent point, we have

$$E^\circ \cup \partial E \subseteq E \cup \partial E \subseteq \bar{E}.$$

To show

$$\bar{E} \subseteq E^\circ \cup \partial E, \tag{3.10}$$

we take any $x_0 \in \bar{E}$, i.e., x_0 is an adherent point of E. If $x_0 \in E^\circ$, we are done. Suppose $x_0 \notin E^\circ$. Then by the definition of an adherent point, x_0 must be a boundary point of E. This proves (3.10). Hence, (3.9) holds.

$\qquad\qquad\qquad\qquad\qquad\qquad\qquad\qquad\qquad\qquad\qquad\qquad\qquad\qquad$ □

Proposition 3.5. *Let (X, d) be a metric space.*

(i) *If $\{E_\lambda\}_{\lambda \in \Lambda}$ is a family of open sets in X, then $\bigcup_{\lambda \in \Lambda} E_\lambda$ is open. If $\{F_\lambda\}_{\lambda \in \Lambda}$ is a family of closed sets, then $\bigcap_{\lambda \in \Lambda} F_\lambda$ is closed.*

(ii) *If $E_1, \cdots, E_n$ are open sets in X, then $E_1 \cap E_2 \cap \cdots \cap E_n$ is open. If $F_1, \cdots, F_n$ are closed, then $F_1 \cup \cdots \cup F_n$ is closed.*

(iii) *For $E \subseteq X$, E° is the largest open set contained in E.*

(iv) *For $E \subseteq X$, $\bar{E}$ is the smallest closed set containing E.*

(v) *For $E \subseteq X$, E is open if and only if E^c is closed.*

(vi) *For $E \subseteq X$, E is closed if and only if for any $\{x_n\} \subseteq E$, $x_n \to \bar{x}$ implies $\bar{x} \in E$.*

The proof is straightforward and is left to the readers.

Note that the intersection of infinitely many open sets might not be open, and the union of infinitely many closed sets might not be closed. Here are counterexamples:

$$\bigcap_{n=1}^{\infty} \left(-\frac{1}{n}, 1 + \frac{1}{n} \right) = [0, 1].$$

$$\bigcup_{n=1}^{\infty} \left[\frac{1}{n}, 1 - \frac{1}{n} \right] = (0, 1).$$

It is clear that the open ball $B(x_0, r)$ is an open set, and its closure

$$\bar{B}(x_0, r) = \{ x \in X \mid d(x, x_0) \leqslant r \}$$

is closed, called the *closed ball* centered at x_0 with radius r.

Example 3.6. (i) Consider $\mathbb{R}^2$ with Euclidean metric ρ_2. Let $X = \{(x, 0) \mid x \in \mathbb{R}\}$. Then (X, ρ_2) is a metric space. Let

$$E = \{(x, 0) \mid -1 < x < 1\} \subseteq X.$$

Then E is open in X, but it is not open in $\mathbb{R}^2$. (Why?)

(ii) Let $\mathbb{R}$ be the metric space with the standard metric ρ. Let $X = (-1, 1)$. Then (X, ρ) is a metric space. Note that X is not closed in $(\mathbb{R}, \rho)$. But, X is always closed in (X, ρ). (Why?)

We now introduce the following definition.

Definition 3.7. Let (X, d) be a metric space and $Y \subseteq X$. Let $E \subseteq Y$. We say that E is *relatively open* (resp. *relatively closed*) in Y if it is open (resp. closed) in the metric space (Y, d).

The following gives a representation of relative open sets and relative closed sets.

Proposition 3.8. *Let (X, d) be a metric space and $E \subseteq Y \subseteq X$.*

(i) *Set E is relatively open (resp. relatively closed) in Y if and only if there exists an open (resp. closed) set $V \subseteq X$ such that $E = V \cap Y$.*

(ii) *For any $E \subseteq Y$, if $\bar{E}$ and $\widehat{E}$ stand for the closures of E in X and in Y, respectively, then*

$$\widehat{E} = \bar{E} \cap Y. \tag{3.11}$$

Proof. (i) Suppose E is relatively open in Y. Then for any $x \in E$, there exists a $\delta(x) > 0$ such that

$$B_Y(x, \delta(x)) \subseteq E.$$

Let

$$V = \bigcup_{x \in E} B_X(x, \delta(x)).$$

This is an open set in X. We now prove that $E = V \cap Y$. First of all, for any $x \in E$,

$$x \in B_Y(x, \delta(x)) \subseteq B_X(x, \delta(x)) \subseteq V,$$

together with $E \subseteq Y$, we obtain

$$E \subseteq V \cap Y.$$

On the other hand, for any $y \in V \cap Y$, since $y \in V$, which implies that for some $x \in E$, $y \in B_X(x, \delta(x))$. Since $y \in Y$ also, one actually has

$$y \in B_X(x, \delta(x)) \cap Y = B_Y(x, \delta(x)) \subseteq E.$$

Hence, we obtain $E = V \cap Y$.

Conversely, suppose $E = V \cap Y$ for some open set V in X. For any $x \in E \subseteq V$, since V is open in X, there exists a $\delta > 0$ such that

$$B_X(x, \delta) \subseteq V.$$

Thus,

$$B_Y(x, \delta) = B_X(x, \delta) \cap Y \subseteq V \cap Y = E.$$

Consequently, x is an interior point of E in (Y, d). Thus, E is open in (Y, d).

Now, let E be relatively closed in Y. Then $Y \setminus E$ is relatively open in Y. Thus, by what we have proved, there exists an open set V in X such that

$$Y \setminus E = V \cap Y.$$

Hence, $V^c = X \setminus V$ is closed in X, and

$$E = Y \setminus (V \cap Y) = Y \cap (V \cap Y)^c = Y \cap (V^c \cup Y^c) = Y \cap V^c.$$

This proves the relatively closed case.

(ii) For any set $E \subseteq Y$, $\bar{E}$ is the smallest closed set in X containing E. Now $\bar{E} \cap Y$ is a closed set in Y containing E. Clearly, it is the smallest one. Hence, (3.11) holds. $\qquad\square$

The following result gives another interesting characterization of the disconnectedness/connectedness, which will be useful later.

Proposition 3.9. *Let (X, d) be a metric space and $E \subseteq X$. Then E is disconnected if and only if there are two disjoint open sets $V_1, V_2 \subseteq X$ such that*

$$E \cap V_1 \neq \varnothing, \quad E \cap V_2 \neq \varnothing, \quad E \subseteq V_1 \cup V_2. \tag{3.12}$$

Note that in the above, we do not assume if E is open or closed.

Proof. Sufficiency. Suppose there are two disjoint open sets $V_1, V_2 \subseteq X$ such that (3.12) holds. Let

$$G_1 = E \cap V_1, \qquad G_2 = E \cap V_2.$$

Consequently,

$$E = E \cap (V_1 \cup V_2) = (E \cap V_1) \cup (E \cap V_2) = G_1 \cup G_2.$$

Further, since $V_1 \cap V_2 = \varnothing$,

$$G_1 \cap G_2 = (E \cap V_1) \cap (E \cap V_2) = E \cap (V_1 \cap V_2) = \varnothing.$$

We claim that $\bar{G}_1 \cap G_2 = \varnothing$. In fact, for any $x \in G_2 = E \cap V_2 \subseteq V_2$, since V_2 is open, there exists a $\delta > 0$ such that

$$B(x, \delta) \subseteq V_2.$$

On the other hand, if this $x \in \bar{G}_1$ as well, then there exists a sequence $x_n \in G_1 = E \cap V_1$ such that $x_n \to x$. Hence, for n large and for the above $\delta > 0$,

$$x_n \in G_1 \cap B(x, \delta) \subseteq E \cap V_1 \cap V_2 \subseteq V_1 \cap V_2 = \varnothing,$$

a contradiction. This shows that $\bar{G}_1 \cap G_2 = \varnothing$. Likewise, $G_1 \cap \bar{G}_2 = \varnothing$. Therefore, E is disconnected.

Necessity. Suppose E is disconnected. Then there are non-empty sets $G_1, G_2 \subseteq X$ such that

$$E = G_1 \cup G_2, \qquad \bar{G}_1 \cap G_2 = \varnothing, \qquad G_1 \cap \bar{G}_2 = \varnothing.$$

For any $x_1 \in G_1$, $x_1 \notin \bar{G}_2$. Thus,

$$d(x_1, G_2) = d(x_1, \bar{G}_2) = \inf_{x_2 \in G_2} d(x_1, x_2) > 0, \qquad \forall x_1 \in G_1.$$

Likewise,

$$d(x_2, G_1) = d(x_2, \bar{G}_1) = \inf_{x_1 \in G_1} d(x_1, x_2) > 0, \qquad \forall x_2 \in G_2.$$

Now, we define

$$V_1 = \bigcup_{x_1 \in G_1} B\left(x_1, \frac{1}{3}d(x_1, G_2)\right), \qquad V_2 = \bigcup_{x_2 \in G_2} B\left(x_2, \frac{1}{3}d(x_2, G_1)\right).$$

Both V_1 and V_2 are open and

$$E \cap V_1 \supseteq G_1, \qquad E \cap V_2 \supseteq G_2, \qquad V_1 \cup V_2 \supseteq G_1 \cup G_2 = E.$$

We now claim that V_1 and V_2 are disjoint. In fact, if there exists an $\bar{x} \in V_1 \cap V_2$, then there are $x_1 \in G_1$ and $x_2 \in G_2$ such that

$$\bar{x} \in B\left(x_1, \frac{1}{3}d(x_1, G_2)\right) \cap B\left(x_2, \frac{1}{3}d(x_2, G_1)\right).$$

Without loss of generality, we assume $d(x_1, G_2) \leqslant d(x_2, G_1)$. Then

$$d(x_2, G_1) \leqslant d(x_1, x_2) \leqslant d(x_1, \bar{x}) + d(x_2, \bar{x})$$
$$\leqslant \frac{1}{3}d(x_1, G_2) + \frac{1}{3}d(x_2, G_1) \leqslant \frac{2}{3}d(x_2, G_1),$$

which is a contradiction. Therefore, V_1 and V_2 are disjoint. $\qquad\square$

Proposition 3.10. *Let (X, d) be a metric space and $E \subseteq X$ be a non-empty set. Then E is disconnected as a set in X if and only if it is disconnected as a metric space, with the induced metric d.*

Proof. As before, for any $G \subseteq E$, we let $\bar{G}$ and $\widehat{G}$ be the closures of G in X and in E, respectively.

Necessity. Suppose E is disconnected as a set in X. Then by definition, there are non-empty sets G_1 and G_2 such that

$$E = G_1 \cup G_2, \qquad \bar{G}_1 \cap G_2 = \varnothing, \qquad G_1 \cap \bar{G}_2 = \varnothing.$$

The above implies that

$$\varnothing = \bar{G}_1 \cap G_2 = \bar{G}_1 \cap G_2 \cap E = (\bar{G}_1 \cap E) \cap G_2 = \widehat{G}_1 \cap G_2,$$

where $\widehat{G}_1 = \bar{G}_1 \cap E$ is the closure of G_1 in E. Likewise,

$$\varnothing = G_1 \cap \bar{G}_2 = G_1 \cap (\bar{G}_2 \cap E) = G_1 \cap \widehat{G}_2,$$

with $\widehat{G}_2 = \bar{G}_2 \cap E$ being the closure of G_2 in E. Hence, by definition, E a is disconnected metric space.

Sufficiency. Suppose E is a disconnected metric space. Then there are non-empty sets $G_1, G_2 \subseteq E$ such that

$$E = G_1 \cup G_2, \qquad \widehat{G}_1 \cap G_2 = \varnothing, \qquad G_1 \cap \widehat{G}_2 = \varnothing.$$

Then, by Proposition 3.8 (ii), we have

$$\widehat{G}_1 = \bar{G}_1 \cap E, \qquad \widehat{G}_2 = \bar{G}_2 \cap E.$$

Consequently, noting $G_1 \cup G_2 = E$,

$$\bar{G}_1 \cap G_2 = \bar{G}_1 \cap E \cap G_2 = \widehat{G}_1 \cap G_2 = \varnothing,$$

and

$$G_1 \cap \bar{G}_2 = G_1 \cap E \cap \bar{G}_2 = G_1 \cap \widehat{G}_2 = \varnothing.$$

Hence, by definition, E is disconnected in X. $\qquad\qquad \square$

The following gives some interesting characterizations of disconnectedness of metric spaces.

Proposition 3.11. *Let (X, d) be a metric space. Then the following are equivalent:*

(i) *X is disconnected.*

(ii) *There are two disjoint open sets $V_1, V_2 \subseteq X$ such that $X = V_1 \cup V_2$.*

(iii) *There are two disjoint closed sets $F_1, F_2 \subseteq X$ such that $X = F_1 \cup F_2$.*

(iv) *There is a proper subset of X which is both open and closed in X.*

Proof. (ii) $\Rightarrow$ (iii). Since $X = V_1 \cup V_2$ with both V_1 and V_2 open and disjoint. Then

$$F_1 = X \setminus V_1 = V_2, \qquad F_2 = X \setminus V_2 = V_1,$$

which are both closed and disjoint, and

$$F_1 \cup F_2 = V_2 \cup V_1 = X.$$

(iii) $\Rightarrow$ (iv). When $X = F_1 \cup F_2$ with both F_1 and F_2 closed and disjoint. Then $F_1^c = F_2$ is open and also closed. This means that X has a proper subset which is both closed and open.

(iv) $\Rightarrow$ (ii). If $G \subseteq X$ is a proper subset of X, which is both open and closed, then G and G^c are two disjoint open sets and

$$X = G \cup G^c.$$

(i) $\Rightarrow$ (iii). By definition, there are disjoint non-empty subsets G_1, G_2 such

$$X = G_1 \cup G_2, \qquad \bar{G}_1 \cap G_2 = \varnothing, \qquad G_1 \cap \bar{G}_2 = \varnothing.$$

Then $G_1 \cap G_2 = \varnothing$. We claim that $G_1 = \bar{G}_1$. In fact, if there exists an $x \in \bar{G}_1 \setminus G_1 \subseteq G_2$, which contradicts $\bar{G}_1 \cap G_2 = \varnothing$. Hence, G_1 is closed. Similarly, G_2 is also closed. Thus, (iii) holds.

(iii) $\Rightarrow$ (i). If $X = G_1 \cup G_2$ with G_1 and G_2 both being closed and disjoint, then

$$\bar{G}_1 \cap G_2 = G_1 \cap G_2 = \varnothing, \qquad G_1 \cap \bar{G}_2 = G_1 \cap G_2 = \varnothing.$$

Hence, X is disconnected. $\qquad\qquad\qquad\qquad\qquad\qquad\qquad\qquad\qquad$ $\square$

Exercises

1. Let (X, d) be a metric space and $E \subseteq X$ be a finite set. Then E is closed.

2. Let (X, d_0) be a discrete metric space. Then any subset E of X is both open and closed.

3. Let (X, d) be a metric space and $E \subseteq X$. Then E° is the largest open set contained in E.

4. Let (X, d) be a metric space and $E \subseteq X$. Then $\bar{E}$ is the smallest closed set containing E.

5. Let (X, d) be a metric space and $E \subseteq X$. Then E is closed if and only if for any convergent sequence $\{x_n\}_{n \geqslant 1} \subseteq E$, the limit $\bar{x}$ of $\{x_n\}_{n \geqslant 1}$ is in E.

6. Let (X_k, d_k) be metric spaces $k = 1, 2, \cdots, m$. Let $X = X_1 \times \cdots \times X_m$ with the product metric

$$d\big((x_1, \cdots, x_m), (y_1, \cdots, y_m)\big) = \sum_{k=1}^{m} d_k(x_k, y_k).$$

Let $E_k \subseteq X_k$, $1 \leqslant k \leqslant m$. If each E_k is open (resp. closed), then $E = E_1 \times \cdots \times E_m$ is open (resp. closed) in (X, d).

7. Show that for any open set $G \subseteq \mathbb{R}$, there exists a finite or an infinite sequence of disjoint open intervals (a_k, b_k), $k \geqslant 1$ such that

$$G = \bigcup_{k \geqslant 1} (a_k, b_k).$$

Extend the above result to the case of $\mathbb{R}^m$.

8. Let (X, d) be a metric space. Suppose $G_1, G_2 \subseteq X$. Then

$$\inf_{x_1 \in G_1, x_2 \in G_2} d(x_1, x_2) = \inf_{x_1 \in \bar{G}_1, x_2 \in \bar{G}_2} d(x_1, x_2).$$

4 Properties of Metric Spaces

We now consider more properties of metric spaces.

Definition 4.1. (i) A set X is said to be *countable* if there exists a *bijection* $f : \mathbb{N} \to X$, i.e., $f(\mathbb{N}) = X$ and $f(\ell) = f(n)$ if and only in $\ell = n$.

(ii) A set X is said to be *at most countable* if it is either finite or countable.

(iii) A set X is said to be *uncountable* if it is not at most countable.

Practically, X is countable if and only if X can be written as follows:

$$X = \{x_1, x_2, x_3, \cdots\}. \tag{4.1}$$

Proposition 4.2. (i) *If X is at most countable, then so is any subset of X.*

(ii) *If X and Y are at most countable, so are $X \cup Y$ and $X \times Y$.*

(iii) *Let X_k, $k = 1, 2, \cdots$ be a sequence of at most countable sets. Then so is $\bigcup_{k=1}^{\infty} X_k$.*

Proof. (i) If X is finite, the conclusion is trivial. Let X be of form (4.1). Let Y be a subset of X. In the case that Y is finite, it is of course at most countable. In the case that Y is infinite, we can find a subsequence $\{x_{\sigma(n)}\}_{n \geqslant 1}$ such that

$$Y = \{x_{\sigma(n)}\}_{n \geqslant 1}.$$

Thus, Y is countable.

(ii) We need only prove the case that

$$X = \{x_1, x_2, x_3, \cdots\}, \quad Y = \{y_1, y_2, y_3, \cdots\}.$$

Clearly,

$$X \cup Y = \{x_1, y_1, x_2, y_2, x_3, y_3, \cdots\}.$$

Hence, $X \cup Y$ is countable. Also, one has

$$X \times Y = \{(x_1, y_1), (x_1, y_2), (x_2, y_1), (x_1, y_3), (x_2, y_2), (x_1, y_3), \cdots\}.$$

This implies that $X \times Y$ is countable.

(iii) Let

$$X_i = \{x_1^i, x_2^i, x_3^i, \cdots\}, \qquad i = 1, 2, 3, \cdots.$$

Then

$$\bigcup_{i=1}^{\infty} X_i = \{x_1^1, x_1^2, x_2^1, x_1^3, x_2^2, x_3^1, \cdots\}.$$

Therefore, $\bigcup_{i=1}^{\infty} X_i$ is countable. $\qquad\qquad\square$

For any $x \in \mathbb{R}$, let $[x]$ be the largest integer that is less than or equal to x. We define the sequence $a_0, a_1, \cdots$ as follows:

$$a_0 = [x] \in \mathbb{Z}, \quad \Rightarrow \quad x - a_0 \in [0, 1);$$

$$a_1 = [10(x - a_0)] \in \{0, 1, \cdots, 9\} \quad \Rightarrow \quad x - a_0 - \frac{a_1}{10} \in \left[0, \frac{1}{10}\right);$$

$$a_2 = \left[10^2\left(x - a_0 - \frac{a_1}{10}\right)\right] \in \{0, 1, \cdots, 9\}$$

$$\Rightarrow \quad x - a_0 - \frac{a_1}{10} - \frac{a_2}{10^2} \in \left[0, \frac{1}{10^2}\right);$$

$$\cdots\cdots\cdots$$

$$a_n = \left[10^n\left(x - \sum_{k=0}^{n-1} \frac{a_k}{10^k}\right)\right] \in \{0, 1, \cdots, 9\} \quad \Rightarrow \quad x - \sum_{k=0}^{n-1} \frac{a_k}{10^k} \in \left[0, \frac{1}{10^n}\right);$$

$$\cdots\cdots\cdots .$$

Hence, for any $x \in \mathbb{R}$, one has

$$x = \sum_{k \geqslant 0} \frac{a_k}{10^k}, \qquad a_0 \in \mathbb{Z}, \ a_k \in \{0, 1, 2, \cdots, 8, 9\}, \ k \geqslant 1. \qquad (4.2)$$

Clearly, for any given real number, the above representation is unique, making use of the convention that any rational number is represented by the corresponding repeating decimal number. Hence, we refer to (4.2) as the *canonical decimal representation* of real numbers.

Proposition 4.3. *For any $a < b$, the set $[a, b]$ is uncountable.*

Proof. Without loss of generality, we consider $[0, 1]$. Suppose $[0, 1]$ is countable. Thus, we let

$$[0, 1] = \{a_1, a_2, \cdots\}.$$

We use the canonical decimal representation for each a_n:

$$a_1 = 0.b_{11}b_{12}b_{13} \cdots,$$
$$a_2 = 0.b_{21}b_{22}b_{23} \cdots,$$
$$a_3 = 0.b_{31}b_{32}b_{33} \cdots,$$
$$\cdots,$$

where $b_{ij} \in \{0, 1, 2, \cdots, 8, 9\}$. Now, we let

$$c = 0.c_1 c_2 c_3 \cdots,$$

with

$$c_1 \in \{0, 1, 2, \cdots, 8, 9\} \setminus \{b_{11}\},$$
$$c_2 \in \{0, 1, 2, \cdots, 8, 9\} \setminus \{b_{22}\},$$
$$c_3 \in \{0, 1, 2, \cdots, 8, 9\} \setminus \{b_{33}\},$$
$$\cdots$$

Then

$$c \notin \{a_1, a_2, a_3, \cdots\} = [0, 1],$$

a contradiction. Hence, $[0, 1]$ is uncountable. □

Definition 4.4. (i) A metric space (X, d) is said to be *complete* if every Cauchy sequence in (X, d) is convergent in (X, d). A normed linear space X is called a *Banach space* if it is complete under the metric induced by the norm. An inner product space is called a *Hilbert space* if it is complete under the metric induced by the inner product.

(ii) A metric space (X, d) is said to be *compact* if every sequence in (X, d) has at least one convergent subsequence.[7] A subset $Y \subseteq X$ is said to

---

[7]Such a compactness definition only works for metric spaces, and it does not work for general topological spaces. This notion is also referred to as the sequential compactness.

be *compact* if subspace (Y, d) is compact. Or, equivalently, for any sequence in Y, there exists a convergent subsequence whose limit is in Y.

(iii) A set Y in a metric space (X, d) is *bounded* if there exists an $x \in X$ and an $r > 0$ such that

$$Y \subseteq B(x, r).$$

A set Y is *totally bounded* if for any $\varepsilon > 0$, there exists an integer $N > 0$ such that

$$Y \subseteq \bigcup_{n=1}^{N} B(x_n, \varepsilon),$$

for some $x_1, x_2, \cdots, x_n \in X$.

Example 4.5. (i) Let (X, d_0) be a discrete metric space. Then it is complete.

(ii) Any finite dimensional normed linear space is complete.

(iii) Let

$$\ell^p = \Big\{ (x_k)_{k \geqslant 1} \mid \sum_{k=1}^{\infty} |x_k|^p < \infty \Big\}, \quad 1 \leqslant p < \infty.$$

We are able to show that ℓ^p is complete. The detailed proof is left to the readers.

Proposition 4.6. *Let (X, d) be a metric space, and (Y, d) be a subspace of (X, d).*

(i) *If (Y, d) is complete, then Y is closed in X.*

(ii) *If (X, d) is complete, and Y is closed in X, then (Y, d) is also complete.*

The proof is straightforward.

The following result is called the *completeness of real numbers*.

Theorem 4.7. *A sequence $\{x_n\} \in \ell^0$ is convergent if and only if it is Cauchy, i.e., $c = \bar{c}$.*

Proof. We already know that if $\{x_n\}$ is convergent, then it is Cauchy (see Proposition 2.2, (ii)).

Conversely, let $\{x_n\}_{n\geqslant 1}$ be Cauchy. Then it is bounded and for any $\varepsilon > 0$, there exists an $N \geqslant 1$ such that

$$|x_k - x_j| < \varepsilon, \qquad \forall k, j \geqslant N.$$

Hence,

$$x_N - \varepsilon \leqslant x_n \leqslant x_N + \varepsilon, \qquad \forall n \geqslant N.$$

Consequently,

$$x_N - \varepsilon \leqslant \varliminf_{n\to\infty} x_n \leqslant \varlimsup_{n\to\infty} x_n \leqslant x_N + \varepsilon.$$

Thus,

$$0 \leqslant \varlimsup_{n\to\infty} x_n - \varliminf_{n\to\infty} x_n \leqslant 2\varepsilon.$$

Since $\varepsilon > 0$ is arbitrary, we obtain that

$$\varliminf_{n\to\infty} x_n = \varlimsup_{n\to\infty} x_n,$$

which leads to the convergence of $\{x_n\}$, by Proposition 2.15. $\qquad\square$

We should point out that to be rigorous, one should go through the theory of the real number system (see Appendix). As a matter of fact, Proposition 2.15 relies on the existence of the supremum and infimum, which is a result of the real number system. Actually, there is a one-to-one correspondence between the set of real numbers and the set of the so-called *equivalent Cauchy sequences*. See the Appendix for details.

For compactness, we have the following simple result.

Proposition 4.8. *Let (X, d) be a metric space.*

(i) *Every finite subset of X, including the empty set, is compact.*

(ii) *If $Y_1, Y_2, \cdots, Y_n$ are compact sets of X, then $Y_1 \cup Y_2 \cup \cdots \cup Y_n$ is also compact.*

(iii) *If $Y_1, Y_2, \cdots, Y_n$ are compact sets of X, then $Y_1 \times Y_2 \times \cdots \times Y_n$ is compact in $X^n \equiv X \times X \times \cdots \times X$.*

Proof. The proofs of (i)–(ii) are obvious. Let us prove (iii). We only prove the case $n = 2$. For any sequence $y_n = (y_n^1, y_n^2) \in Y_1 \times Y_2$, by the compactness of Y_1, we have a convergent subsequence $\{y_{\sigma_1(n)}^1\}_{n\geqslant 1}$ of $\{y_n^1\}_{n\geqslant 1}$ and some $\bar{y}^1 \in Y_1$ such that

$$d(y_{\sigma_1(n)}^1, \bar{y}^1) \to 0, \qquad n \to \infty.$$

Next, by the compactness of Y_2, there exists a subsequence $\{y^2_{\sigma_2(\sigma_1(n))}\}_{n \geqslant 1}$ of $\{y^2_{\sigma_1(n)}\}_{n \geqslant 1}$ and a $\bar{y}^2 \in Y_2$ such that

$$d(y^2_{\sigma_2(\sigma_1(n))}, \bar{y}^2) \to 0, \qquad n \to \infty.$$

Consequently, $(\bar{y}^1, \bar{y}^2) \in Y_1 \times Y_2$ and

$$d((y^1_{\sigma_2(\sigma_1(n))}, y^2_{\sigma_2(\sigma_1(n))}), (\bar{y}^1, \bar{y}^2)) \to 0, \qquad n \to \infty.$$

This proves the compactness of $Y_1 \times Y_2$. $\qquad\square$

The following result gives deeper property of compact sets.

Proposition 4.9. *Let (X, d) be a compact metric space. Then X is complete and bounded; and any $Y \subseteq X$ is compact if and only if it is closed.*

Proof. Suppose (X, d) is compact. If (X, d) is not bounded, then pick any $x_1 \in X$, and we can find

$$x_2 \in X \setminus B(x_1, 1).$$

Next, we can find

$$x_3 \in X \setminus B(x_1, 1) \cup B(x_2, 1),$$

since $B(x_1, 1) \cup B(x_2, 1)$ is a bounded set. We continue this procedure and obtain a sequence x_n with

$$x_n \in X \setminus \bigcup_{k=1}^{n-1} B(x_k, 1).$$

Then we see that

$$d(x_i, x_k) \geqslant 1, \qquad \forall i \neq j.$$

Hence, this sequence does not have a convergent subsequence, a contradiction, proving the boundedness of X.

Next, let $\{x_n\}_{n \geqslant 1}$ be a Cauchy sequence in X. By the compactness of X, there exists a subsequence $\{x_{n_k}\}_{k \geqslant 1}$ converging to some limit point $\bar{x} \in X$. Then we can show that the whole sequence is convergent to the same limit $\bar{x}$. Hence, X is complete.

The conclusion for Y is clear. $\qquad\square$

Note that the converse of the above is not true, i.e., bounded and closed set is not necessarily compact. For example, if (X, d_0) is a discrete metric space and $Y \subseteq X$ is an infinite set. Then Y is bounded and closed, but not compact. However, we have the following positive result.

Theorem 4.10. (Bolzano-Weierstrass) *Let $\{x_n\}_{n \geqslant 1}$ be a bounded sequence in $\mathbb{R}^m$ with the ℓ_p-metric ρ_p. Then it has a convergent subsequence.*

Proof. We only prove the case $m = 1$. Since $\{x_n\}_{n \geqslant 1}$ is bounded, $L = \varlimsup_{n \to \infty} x_n$ exists. Then by Proposition 2.17, this L is a limit point of $\{x_n\}_{n \geqslant 1}$, i.e., for some subsequence $x_{n_k} \to L$. $\square$

Corollary 4.11. (Heine–Borel) *Let $\mathbb{R}^m$ be the metric space with the ℓ^p-metric ρ_p for some $1 \leqslant p \leqslant \infty$. Let $E \subseteq \mathbb{R}^m$. Then E is compact if and only if E is bounded and closed.*

Proof. $\Rightarrow$ It follows from Proposition 4.9.

$\Leftarrow$ It follows from Bolzano-Weierstrass Theorem (Theorem 4.10). $\square$

Theorem 4.12. (Finite Open Cover) *Let (X, d) be a metric space and $Y \subseteq X$. Then Y is compact if and only if it has the finite open cover property: For any family of open sets $\{V_\lambda \mid \lambda \in \Lambda\}$ in X such that*

$$Y \subseteq \bigcup_{\lambda \in \Lambda} V_\lambda,$$

there exists a finite subset $\Lambda_0 \subseteq \Lambda$ such that

$$Y \subseteq \bigcup_{\lambda \in \Lambda_0} V_\lambda.$$

Proof. Necessity. Suppose Y is compact and $\{V_\lambda \mid \lambda \in \Lambda\}$ is an open cover of Y. Then for each $y \in Y$, there exists a $\lambda \in \Lambda$ such that $y \in V_\lambda$. Since V_λ is open, there exists a $\delta > 0$ such that $B(y, \delta) \subseteq V_\lambda$. Now, we let

$$\delta(y) = \sup\{\delta > 0 \mid B(y, \delta) \subseteq V_\lambda, \text{ for some } \lambda \in \Lambda\} > 0.$$

Define

$$\delta_0 = \inf\{\delta(y) \mid y \in Y\} \geqslant 0.$$

We have two cases.

Case 1. $\delta_0 = 0$. Then for each $n \geqslant 1$, we can find a $y_n \in Y$ such that

$$\delta(y_n) < \frac{1}{n}.$$

Since Y is compact, there exists a subsequence y_{n_k} converging to some $\bar{y} \in Y$. Then for some $\lambda_0 \in \Lambda$, we have a $\bar{\delta} \in (0, \delta(\bar{y}))$ such that

$$\bar{y} \in B(\bar{y}, \bar{\delta}) \subseteq V_{\lambda_0}.$$

Since $\{y_{n_k}\}$ converges to $\bar{y}$, there exists a $K > 0$ such that

$$d(y_{n_k}, \bar{y}) < \frac{\bar{\delta}}{2}, \qquad \forall k \geqslant K.$$

Consequently, for any $z \in B(y_{n_k}, \frac{\bar{\delta}}{2})$, we have

$$d(z, \bar{y}) \leqslant d(z, y_{n_k}) + d(y_{n_k}, \bar{y}) < \frac{\bar{\delta}}{2} + \frac{\bar{\delta}}{2} = \bar{\delta}.$$

This means that

$$B(y_{n_k}, \frac{\bar{\delta}}{2}) \subseteq B(\bar{y}, \bar{\delta}) \subseteq V_{\lambda_0}.$$

Hence,

$$\frac{\bar{\delta}}{2} \leqslant \delta(y_{n_k}) < \frac{1}{n_k} \to 0,$$

a contradiction, which means that this case will not happen.

Case 2. $\delta_0 > 0$. Then we must have

$$\delta(y) > \frac{\delta_0}{2}, \qquad \forall y \in Y.$$

This implies that for each $y \in Y$, there exists a $\lambda(y) \in \Lambda$ such that

$$B(y, \frac{\delta_0}{2}) \subseteq V_{\lambda(y)}.$$

Now, we construct a sequence. Pick any $y_1 \in Y$. If

$$Y \subseteq V_{\lambda(y_1)},$$

we are done. Suppose the above is not true. Then we can find

$$y_2 \in Y \setminus V_{\lambda(y_1)} \subseteq Y \setminus B(y_1, \frac{\delta_0}{2}).$$

Clearly,

$$d(y_2, y_1) \geqslant \frac{\delta_0}{2}.$$

If

$$Y \subseteq V_{\lambda(y_1)} \cup V_{\lambda(y_2)}.$$

we are done. Otherwise, we can find

$$y_3 \in Y \setminus \left(V_{\lambda(y_1)} \cup V_{\lambda(y_2)} \right) \subseteq Y \setminus \left(B(y_1, \frac{\delta_0}{2}) \cup B(y_2, \frac{\delta_0}{2}) \right).$$

Thus,

$$d(y_3, y_1), d(y_3, y_2) \geqslant \frac{\delta_0}{2}.$$

If the above procedure stops at a finite step, we are done. Otherwise, a sequence y_n can be constructed so that

$$y_n \in Y \setminus \left(\bigcup_{k=1}^{n-1} V_{\lambda(y_k)} \right) \subseteq Y \setminus \left(\bigcup_{k=1}^{n-1} B(y_k, \frac{\delta_0}{2}) \right).$$

Hence,

$$d(y_n, y_k) \geqslant \frac{\delta_0}{2}, \qquad \forall 1 \leqslant k < n.$$

This implies that Y contains a sequence $\{y_n\}_{n \geqslant 1}$ which does not have a convergent subsequence, a contradiction.

Sufficiency. By contradiction. Suppose Y is not compact. Then there exists a sequence $\{y_n\}_{n \geqslant 1} \subseteq Y$, which does not have convergent subsequences. In particular, $\{y_n\}_{n \geqslant 1}$ is an infinite set. We claim that for any $y \in Y$, there exists an $\varepsilon = \varepsilon(y) > 0$ such that

$$B(y, \varepsilon(y)) \cap \{y_n \mid n \geqslant 1\}$$

is a finite set. In fact, otherwise, by shrinking $\varepsilon > 0$, we will obtain a subsequence of $\{y_n\}_{n \geqslant 1}$ converging to y. Now, we have that

$$Y \subseteq \bigcup_{y \in Y} B(y, \varepsilon(y)).$$

Then by assumption, there is a finite open cover, i.e., there are $\bar{y}_1, \cdots, \bar{y}_m \in Y$ such that

$$\{y_n\}_{n \geqslant 1} \subseteq Y \subseteq \bigcup_{k=1}^{m} B(\bar{y}_k, \varepsilon(\bar{y}_k)).$$

This leads to a contradiction since each $B(\bar{y}_k, \varepsilon(\bar{y}_k))$ contains only finitely many points of $\{y_n\}_{n \geqslant 1}$. $\qquad \square$

Theorem 4.13. (Finite Intersection) *Let (X, d) be a metric space and $Y \subseteq X$. Then Y is compact if and only if it has the finite intersection property: If $\{F_\lambda : \lambda \in \Lambda\}$ is a family of closed sets in X such that for any finite set $\Lambda_0 \subseteq \Lambda$,*

$$Y \cap \left(\bigcap_{\lambda \in \Lambda_0} F_\lambda \right) \neq \varnothing, \tag{4.3}$$

then

$$Y \cap \left(\bigcap_{\lambda \in \Lambda} F_\lambda \right) \neq \varnothing. \tag{4.4}$$

Proof. Necessity. Suppose Y is compact and $\{F_\lambda \mid \lambda \in \Lambda\}$ has the property that for any finite set $\Lambda_0 \subseteq \Lambda$, (4.3) holds, but

$$Y \cap \Big(\bigcap_{\lambda \in \Lambda} F_\lambda \Big) = \varnothing.$$

Then

$$Y \subseteq \Big(\bigcap_{\lambda \in \Lambda} F_\lambda \Big)^c = \bigcup_{\lambda \in \Lambda} F_\lambda^c.$$

Thus, $\{F_\lambda^c \mid \lambda \in \Lambda\}$ is an open cover of Y. Since Y is compact, there exists a finite open cover, i.e., there exists a finite set $\Lambda_0 \subseteq \Lambda$ such that

$$Y \subseteq \bigcup_{\lambda \in \Lambda_0} F_\lambda^c = \Big(\bigcap_{\lambda \in \Lambda_0} F_\lambda \Big)^c,$$

which implies that

$$Y \cap \Big(\bigcap_{\lambda \in \Lambda_0} F_\lambda \Big) = \varnothing.$$

This is a contradiction.

Sufficiency. Suppose $\{V_\lambda \mid \lambda \in \Lambda\}$ is an open cover of Y. Then we claim that it must have a finite open cover of Y. If not, i.e., for any finite set $\Lambda_0 \subseteq \Lambda$, one has

$$\varnothing \neq Y \cap \Big(\bigcup_{\lambda \in \Lambda_0} V_\lambda \Big)^c = Y \cap \Big(\bigcap_{\lambda \in \Lambda_0} V_\lambda^c \Big).$$

Since each V_λ^c is closed, by assumption, we must have

$$\varnothing \neq Y \cap \Big(\bigcap_{\lambda \in \Lambda} V_\lambda^c \Big) = Y \cap \Big(\bigcup_{\lambda \in \Lambda} V_\lambda \Big)^c,$$

which means that $\{V_\lambda \mid \lambda \in \Lambda\}$ is not an open cover of Y, a contradiction. $\quad\square$

Theorem 4.14. (Nested Compact Sets) *Let* (X, d) *be a metric space and* K_n *be a sequence of nested non-empty compact sets:*

$$K_1 \supseteq K_2 \supseteq K_3 \supseteq \cdots.$$

Then

$$\bigcap_{n=1}^{\infty} K_n \neq \varnothing. \tag{4.5}$$

Moreover, if

$$\operatorname{diam}(K_n) \equiv \sup_{x,y \in K_n} d(x, y) \to 0, \qquad n \to \infty, \tag{4.6}$$

then there exists a unique $\bar{x} \in X$ *such that*

$$\bigcap_{n=1}^{\infty} K_n = \{\bar{x}\}. \tag{4.7}$$

Proof. Define

$$V_n = K_n^c, \qquad n \geqslant 1.$$

Then $\{V_n \mid n \geqslant 1\}$ is a sequence of open sets in X. Now, if (4.5) is not true, then

$$\bigcup_{n \geqslant 1} V_n = \bigcup_{n \geqslant 1} K_n^c = \left(\bigcap_{n \geqslant 1} K_n \right)^c = X \supseteq K_1.$$

By the compactness of K_1, there exists an $N \geqslant 1$ such that

$$K_1 \subseteq \bigcup_{n=1}^{N} V_n = \bigcup_{n=1}^{N} K_n^c = \left(\bigcap_{n=1}^{N} K_n \right)^c = K_N^c,$$

which implies

$$\varnothing \neq K_N = K_N \cap K_1 \subseteq K_N \cap K_N^c = \varnothing,$$

a contradiction. Hence, (4.5) holds.

Now, suppose (4.6) holds, then for any $\bar{x}, \widehat{x} \in \bigcap_{n \geqslant 1} K_n$, we have

$$d(\bar{x}, \widehat{x}) \leqslant \operatorname{diam}(K_n) \to 0, \qquad n \to \infty.$$

Hence, (4.7) holds. $\qquad \qquad \square$

Next result is relevant to the total boundedness.

Theorem 4.15. *Let (X, d) be a metric space and $Y \subseteq X$. Then Y is compact if and only if Y is complete and totally bounded.*

Proof. Necessity. Suppose Y is compact. Then Y has to be complete. Next, for any $\varepsilon > 0$, we have

$$Y \subseteq \bigcup_{x \in Y} B(x, \varepsilon).$$

Then by finite open cover theorem, we see that Y is totally bounded.

Sufficiency. Let $\{y_n\}_{n \geqslant 1} \subseteq Y$ be a sequence. For $k = 1$, Y is covered by finitely many open balls of radius 1. Among them, there is a ball, denoted by $B(x_1, 1)$, containing a subsequence of $\{y_n\}_{n \geqslant 1}$. We denote it by

$$y_{n,1} \in B(x_1, 1) \cap Y, \qquad \forall n \geqslant 1.$$

Next, since $\bar{B}(x_1, 1) \cap Y$ is compact, it is covered by finitely many balls of radius $\frac{1}{2}$. Then there exists an $x_2 \in B(x_1, 1)$ and a subsequence $\{y_{n,2}\}_{n \geqslant 1}$ of $\{y_{n,1}\}_{n \geqslant 1}$ such that

$$y_{n,2} \in B\left(x_2, \frac{1}{2}\right) \cap Y, \qquad n \geqslant 1.$$

Continue this procedure, we can find $\{x_k\}_{k \geqslant 1} \subseteq X$ and subsequences $\{y_{n,k}\}_{n \geqslant 1}$ such that

$$x_k \in B\left(x_{k-1}, \frac{1}{2^{k-1}}\right), \quad y_{n,k} \in B\left(x_k, \frac{1}{2^k}\right) \cap Y, \qquad n \geqslant 1.$$

Now, we let $z_k = y_{k,k}$. Then $\{z_k\}_{k \geqslant 1}$ is a subsequence of $\{y_n\}_{n \geqslant 1}$ such that

$$z_k \in B\left(x_k, \frac{1}{2^k}\right) \cap Y, \qquad k \geqslant 1. \tag{4.8}$$

Note that,

$$d(x_i, x_{i+j}) \leqslant \sum_{k=0}^{j-1} d(x_{i+k}, x_{i+k+1}) < \sum_{k=0}^{j-1} \frac{1}{2^{i+k}} < \frac{1}{2^{i-1}}.$$

This means that $\{x_k\}_{k \geqslant 1}$ is a Cauchy sequence. By (4.8) we see that

$$\lim_{k \to \infty} d(x_k, z_k) = 0.$$

Hence, $\{z_k\}_{k \geqslant 1}$ is also Cauchy. Then by the completeness of Y, one has the convergence of $\{z_k\}_{k \geqslant 1}$. This proves the compactness of Y. $\qquad\square$

The following example shows that in an infinite dimensional space, bounded closed set is not necessarily compact, in general.

Example 4.16. Recall ℓ^p ($p \in [1, \infty)$). Look at the closed unit ball centered at 0:

$$\bar{B}(0,1) = \left\{ (x^k) \in \ell^p \mid \sum_{k=1}^{\infty} |x^k|^p \leqslant 1 \right\}.$$

It is bounded and closed. We claim that it is not compact. In fact, let us look at the following sequence:

$$x_n = (\delta_{nk}), \qquad \forall n \geqslant 1,$$

where

$$\delta_{nk} = \begin{cases} 1, & k = n, \\ 0, & k \neq n. \end{cases}$$

Then for any $n \neq m$, one has

$$\rho_p(x_n, x_m) = 2^{\frac{1}{p}}.$$

Hence, the sequence $\{x_n\}_{n \geqslant 1}$ does not have a convergent subsequence. This proves our claim.

Definition 4.17. Let (X, d) be a metric space. Then it is said to be *separable* if there exists a countable set X_0 such that

$$\bar{X}_0 = X.$$

In other words, X admits a countable dense subset.

Proposition 4.18. *Let (X, d) be a compact metric space. Then X is separable.*

Proof. Since X is compact, it is totally bounded. Consequently, for any $n \geqslant 1$, there exists a finite set $\{x_k^n \mid 1 \leqslant k \leqslant K_n\}$ such that

$$X \subseteq \bigcup_{k=1}^{K_n} B\Big(x_k^n, \frac{1}{n}\Big).$$

Now, let

$$X_0 = \Big\{x_k^n \mid 1 \leqslant k \leqslant K_n, \ n \geqslant 1\Big\}.$$

Then X_0 is countable and it is dense in X. Hence, X is separable. $\qquad\square$

It is clear that if (X, d) is separable, then any subset $Y \subseteq X$ is also separable.

Example 4.19. (i) If (X_k, d_k) $(1 \leqslant k \leqslant n)$ is separable, then $X_1 \times X_2 \times \cdots \times X_n$ is separable under the corresponding product metric.

(ii) Any subset Y of $\mathbb{R}^n$ is separable under the usual product metric of ℓ^p-metric $(1 \leqslant p \leqslant \infty)$.

(iii) Let (X, d_0) be a metric space with the discrete metric, and X be an uncountable set. Then (X, d_0) is not separable.

Exercises

1. Show that the set $\mathbb{Q}$ of all rational numbers is countable and the set $\mathbb{R} \setminus \mathbb{Q}$ of all irrational numbers is uncountable.

2. Prove the Bolzano–Weierstrass Theorem (Any bounded sequence in $\mathbb{R}^m$ contains a convergent subsequence) by the Nested Compact Sets Theorem.

3. Prove the Heine–Borel Theorem by the Nested Compact Sets Theorem.

4. Let $S = \{\frac{1}{n} \mid n \in \mathbb{N}\} \cup \{0\}$. Then S is closed.

5. Show that any discrete metric space (X, d_0) must be complete.

6. Let (X, d) be a complete metric space and $A_n \subseteq X$ satisfy

$$A_{n+1} \subseteq A_n, \qquad \forall n \geqslant 1,$$
$$\operatorname{diam}(A_n) \equiv \sup_{x,y \in A_n} d(x, y) \to 0, \qquad n \to \infty.$$

Let $a_n \in A_n$. Then the sequence $\{a_n\}$ converges. Find the difference between this result and the Nested Compact Sets Theorem.

7. Show that $\mathbb{Q} \cap [0, 1]$ is totally bounded.

8. Show that ℓ^p is separable for $p \in [1, \infty)$ and ℓ^∞ is not separable.

9. Let (X, d_0) be a discrete metric space. Let $E \subseteq X$ be a non-empty subset. Then E is compact if and only if E is a finite set.

10. Let (X, d) be a metric space.

(i) Suppose $K_1, \cdots, K_n \subseteq X$ are compact sets. Then $\bigcup_{i=1}^{n} K_i$ is compact.

(ii) Suppose $\{K_\lambda, \lambda \in \Lambda\}$ is a family of compact subsets in X. Then $\bigcap_{\lambda \in \Lambda} K_\lambda$ is compact.

11. Let K_1 and K_2 be two disjoint compact sets in a metric space (X, d). Then

$$a = \inf_{x_x \in K_1, x_2 \in K_2} d(x_1, x_2) > 0.$$

Moreover there are $\bar{x}_1 \in K_1$ and $\bar{x}_2 \in K_2$ such that

$$a = d(\bar{x}_1, \bar{x}_2).$$

What happens if we only assume K_1 and K_2 to be closed?

12. Let $\{x_n\} \in \ell^\infty$. Then the set $\mathcal{L}$ of all limit points of $\{x_n\}$ is non-empty. Moreover, for any sequence $\{y_n\} \subseteq \mathcal{L}$, if $\{y_n\}$ is convergent to $\bar{y}$, then $\bar{y} \in \mathcal{L}$.

Chapter 2

Functions on Metric Spaces

This chapter is devoted to the study of functions on metric spaces.

1 Continuity

We first give the following definition.

Definition 1.1. Let (X, d_X) and (Y, d_Y) be two metric spaces.

(i) Let $E \subseteq X$ and $f : E \to Y$. Let x_0 be a limit point of E. We say that $f(x)$ is convergent to $L \in Y$ as $x \to x_0$ if for any $\varepsilon > 0$, there exists a $\delta > 0$ such that

$$d_Y(f(x), L) < \varepsilon, \qquad \forall x \in E, \ 0 < d_X(x, x_0) < \delta.$$

We denote the above by

$$\lim_{x \to x_0} f(x) \equiv \lim_{E \ni x \to x_0} f(x) = L.$$

(ii) Let $f : X \to Y$. We say that $f(\cdot)$ is *continuous* at $x_0 \in X$ if for every $\varepsilon > 0$, there exists a $\delta > 0$ such that

$$d_Y\big(f(x), f(x_0)\big) < \varepsilon, \qquad \forall d_X(x, x_0) < \delta.$$

We say that $f(\cdot)$ is continuous on X if it is continuous at every $x \in X$.

The most important example of the above is the case $(X, d_X) = (\mathbb{R}^m, \rho_2)$ and $(Y, d_Y) = (\mathbb{R}, \rho)$. Any $f : \mathbb{R}^m \to \mathbb{R}$, with $m > 1$, is called a *multivariable function*. The following gives some equivalent conditions for a function being continuous at a given point x_0.

Theorem 1.2. Let (X, d_X) and (Y, d_Y) be metric spaces. Let $f : X \to Y$ and $x_0 \in X$. Then the following are equivalent:

(i) $f(\cdot)$ *is continuous at* x_0.

(ii) *Whenever* $\{x_n\}_{n \geqslant 1} \subseteq X$ *is convergent to* x_0 *under* d_X, *the sequence* $\{f(x_n)\}_{n \geqslant 1}$ *is convergent to* $f(x_0)$ *under* d_Y.

(iii) *For every open set* $V \subseteq Y$ *containing* $f(x_0)$, *there exists an open set* $U \subseteq X$ *containing* x_0 *such that*

$$f(U) \subseteq V.$$

The proof of the above result is straightforward. Such a result leads to the following which is about the continuity of a function $f : X \to Y$ on X.

Theorem 1.3. *Let* (X, d_X) *and* (Y, d_Y) *be metric spaces. Let* $f : X \to Y$. *Then the following are equivalent:*

(i) $f(\cdot)$ *is continuous on* X.

(ii) *Whenever* $\{x_n\} \subseteq X$ *is convergent to some* x_0 *under* d_X, *the sequence* $\{f(x_n)\}$ *is convergent to* $f(x_0)$ *under* d_Y.

(iii) *For every open set* $V \subseteq Y$, $f^{-1}(V)$ *is open in* X, *where*

$$f^{-1}(V) = \{x \in X \mid f(x) \in V\}.$$

(iv) *For every closed set* F *in* Y, $f^{-1}(F)$ *is closed in* X.

Proof. (i) $\Rightarrow$ (ii). Let $x_n \in X$ with $x_n \to x_0$. By the continuity of $f(\cdot)$, for any $\varepsilon > 0$, there exists a $\delta > 0$ such that

$$d_Y\big(f(x), f(x_0)\big) < \varepsilon, \qquad d_X(x, x_0) < \delta.$$

Also, there exists an $N \geqslant 1$ such that

$$d_X(x_n, x_0) < \delta, \qquad n \geqslant N.$$

This leads to

$$d_Y\big(f(x_n), f(x_0)\big) < \varepsilon, \qquad n \geqslant N.$$

Hence, (ii) holds.

(ii) $\Rightarrow$ (iii). Let $V \subseteq Y$ be an open set. If $f^{-1}(V)$ is not open, then there exists an $x_0 \in f^{-1}(V)$ and a sequence $x_n \in f^{-1}(V)^c$ converging to x_0. Hence $f(x_n) \to f(x_0)$. Since V is open, there exists an $\varepsilon > 0$ such that

$$B_Y(f(x_0), \varepsilon) \subseteq V.$$

Therefore, there exists an $N \geqslant 1$ such that

$$f(x_n) \in B_Y\big(f(x_0), \varepsilon\big), \qquad \forall n \geqslant N.$$

This implies that

$$x_n \in f^{-1}\big(B_Y(f(x_0), \varepsilon)\big) \subseteq f^{-1}(V),$$

a contradiction.

(iii) $\Rightarrow$ (iv). For any closed set $F \subseteq Y$, F^c is open. Thus,

$$f^{-1}(F^c) = f^{-1}(F)^c,$$

is open, leading to that $f^{-1}(F)$ is closed.

(iv) $\Rightarrow$ (iii). The proof is the same as (iii) $\Rightarrow$ (iv).

(iii) $\Rightarrow$ (i). Let $x_0 \in X$. For any $\varepsilon > 0$, $B_Y(f(x_0), \varepsilon)$ is open. Hence $f^{-1}\big(B_Y(f(x_0), \varepsilon)\big)$ is open, and

$$x_0 \in f^{-1}\big(B_Y(f(x_0), \varepsilon)\big).$$

Thus, there exists a $\delta > 0$ such that

$$B_X(x_0, \delta) \subseteq f^{-1}\big(B_Y(f(x_0), \varepsilon)\big).$$

This means

$$f\big(B_X(x_0, \delta)\big) \subseteq B_Y(f(x_0), \varepsilon),$$

i.e.,

$$d_Y(f(x), f(x_0)) < \varepsilon, \qquad \forall d_X(x, x_0) < \delta.$$

This completes the proof. $\qquad\qquad\qquad\qquad\qquad\qquad\qquad\qquad$ $\square$

Proposition 1.4. *Let (X, d_X), (Y, d_Y), and (Z, d_Z) be metric spaces. Let $f : X \to Y$, $g : Y \to Z$. If $f(\cdot)$ is continuous at $x_0 \in X$ and $g(\cdot)$ is continuous at $f(x_0)$, then $g \circ f : X \to Z$ is continuous at x_0. Consequently, if $f(\cdot)$ is continuous on X and $g(\cdot)$ is continuous on Y, then $(g \circ f)(\cdot)$ is continuous on X.*

The proof is straightforward.

Now, we look at an important special case $(Y, d_Y) = (\mathbb{R}, \rho)$. Note that in $\mathbb{R}$, there is an additional structure — the order structure. Due to such a structure, we can say a little more on any bounded, not necessarily continuous function $f : X \to \mathbb{R}$. Suppose $x_0 \in X$ and $f(\cdot)$ is bounded in a ball $B(x_0, r)$, for some $r > 0$. For any $\delta \in (0, r]$, we have the existence of

$$f^-(x_0; \delta) = \inf_{x \in B(x_0, \delta) \setminus \{x_0\}} f(x), \qquad f^+(x_0; \delta) = \sup_{x \in B(x_0, \delta) \setminus \{x_0\}} f(x).$$

Clearly, $\delta \mapsto f^-(x_0; \delta)$ is bounded and monotone non-increasing. Thus, the following limit exists:

$$\varliminf_{x \to x_0} f(x) \equiv \lim_{\delta \downarrow 0} f^-(x_0; \delta) = \sup_{\delta > 0} \inf_{x \in B(x_0, \delta) \setminus \{x_0\}} f(x).$$

Likewise, the following is well-defined:

$$\varlimsup_{x \to x_0} f(x) \equiv \lim_{\delta \downarrow 0} f^+(x_0; \delta) = \inf_{\delta > 0} \sup_{x \in B(x_0, \delta) \setminus \{x_0\}} f(x).$$

We called the above the *limit inferior* (or simply liminf) and *limit superior* (or limsup) of $f(\cdot)$ as $x \to x_0$. Then we can introduce the following definition.

Definition 1.5. Let (X, d) be a metric space and $f : X \to \mathbb{R}$ be a bounded function.

(i) We say that $f(\cdot)$ is *lower semi-continuous* at x_0 if

$$f(x_0) \leqslant \varliminf_{x \to x_0} f(x). \tag{1.1}$$

(ii) We say that $f(\cdot)$ is *upper semi-continuous* at x_0 if

$$f(x_0) \geqslant \varlimsup_{x \to x_0} f(x). \tag{1.2}$$

(iii) If $f(\cdot)$ is lower (resp. upper) semi-continuous at every point in X, we say that $f(\cdot)$ is lower (resp. upper) semi-continuous on X.

To get some feeling about the lower and upper semi-continuous functions, let us present the following example.

Example 1.6. Let

$$\varphi(x) = \begin{cases} 1, & x = 0, \\ 0, & x \neq 0; \end{cases} \qquad \psi(x) = \begin{cases} -1, & x = 0, \\ 0, & x \neq 0. \end{cases}$$

Then $\varphi(\cdot)$ is upper semi-continuous, and $\psi(\cdot)$ is lower semi-continuous.

The following gives a characterization of lower semi-continuous functions.

Proposition 1.7. *Let (X, d) be a metric space and $f : X \to \mathbb{R}$ be a bounded function. Then the following are equivalent:*

(i) *$f(\cdot)$ is lower semi-continuous on X.*

(ii) *For every $x_0 \in X$, for any $\varepsilon > 0$, there exists a $\delta > 0$ such that*

$$f(x) > f(x_0) - \varepsilon, \qquad \forall x \in B(x_0, \delta).$$

(iii) *For every $a \in \mathbb{R}$, the set*

$$(f > a) \equiv \{x \in X \mid f(x) > a\}$$

is open.

Proof. (i) $\Rightarrow$ (ii). By the lower semi-continuity of $f(\cdot)$ at x_0, we have

$$f(x_0) \leqslant \varliminf_{x \to x_0} f(x) = \lim_{\delta \downarrow 0} \inf_{x \in B(x_0, \delta) \setminus \{x_0\}} f(x).$$

Then for any $\varepsilon > 0$, there exists a $\delta > 0$ such that

$$f(x_0) < \inf_{x \in B(x_0, \delta) \setminus \{x_0\}} f(x) + \varepsilon \leqslant f(x) + \varepsilon, \qquad \forall x \in B(x_0, \delta) \setminus \{x_0\}.$$

The above is true, of course, for $x = x_0$. Thus, (ii) holds.

(ii) $\Rightarrow$ (iii). Fix any $a \in \mathbb{R}$. For any $x_0 \in (f > a)$, we have

$$f(x_0) > a.$$

Then by (ii), for $\varepsilon = \frac{f(x_0) - a}{2}$, there exists a $\delta > 0$ such that

$$f(x) > f(x_0) - \varepsilon = f(x_0) - \frac{f(x_0) - a}{2} = \frac{f(x_0) + a}{2} > a,$$
$$\forall x \in B(x_0, \delta).$$

This implies that

$$B(x_0, \delta) \subseteq (f > a).$$

Thus, $(f > a)$ is open; (iii) holds.

(iii) $\Rightarrow$ (i). For any $x_0 \in X$ and any $\varepsilon > 0$,

$$x_0 \in (f > f(x_0) - \varepsilon).$$

Then there exists a $\delta > 0$ such that

$$f(x) > f(x_0) - \varepsilon, \qquad \forall x \in B(x_0, \delta).$$

Hence,

$$\varliminf_{x \to x_0} f(x) \geqslant f(x_0) - \varepsilon,$$

for any $\varepsilon > 0$, which leads to the lower semi-continuity of $f(\cdot)$ at x_0. $\square$

Likewise, we have the following result for the upper semi-continuous functions.

Proposition 1.8. *Let (X, d) be a metric space and $f : X \to \mathbb{R}$ be a bounded function. Then the following are equivalent:*

(i) *$f(\cdot)$ is upper semi-continuous.*

(ii) *For every $x_0 \in X$, for any $\varepsilon > 0$, there exists a $\delta > 0$ such that*

$$f(x) < f(x_0) + \varepsilon, \qquad \forall x \in B(x_0, \delta).$$

(iii) *For every $a \in \mathbb{R}$, the set*

$$(f < a) \equiv \{x \in X \mid f(x) < a\}$$

is open.

The following is pretty obvious.

Corollary 1.9. *Let (X, d) be a metric space and $f : X \to \mathbb{R}$ be a bounded function. Then $f(\cdot)$ is continuous if and only if it is both lower and upper semi-continuous.*

Next, we look at two interesting functions.

Example 1.10. (i) The following is called the *Dirichlet function*:

$$f(x) = \begin{cases} 1, & x \in \mathbb{Q}, \\ 0, & x \in \mathbb{R} \setminus \mathbb{Q}. \end{cases}$$

Since for any $\bar{x} \in \mathbb{R}$,

$$\overline{\lim_{x \to \bar{x}}} f(x) = 1, \qquad \underline{\lim_{x \to \bar{x}}} f(x) = 0,$$

we see that $\lim_{x \to \bar{x}} f(x)$ fails to exist for any $\bar{x} \in \mathbb{R}$, which implies that the Dirichlet function is nowhere continuous. It is upper semi-continuous at every rational point, and lower semi-continuous at every irrational point.

(ii) The following is called the *Riemann function*:

$$f(x) = \begin{cases} \dfrac{1}{q}, & \text{if } x = \frac{p}{q} \in \mathbb{Q}, \text{ with } q > 0 \text{ and } p \text{ and } q \text{ are co-prime,} \\ 0, & \text{if } x \in \mathbb{R} \setminus \mathbb{Q}. \end{cases}$$

Since for any $\bar{x} \in \mathbb{R}$,

$$\lim_{x \to \bar{x}} f(x) = 0, \qquad \text{(why?)}$$

we see that this function is continuous at every irrational point and is discontinuous at every rational point.

We now introduce the following definition which is concerned with functions from a metric space to itself.

Definition 1.11. *Let (X, d) be a metric space, and $f : X \to X$. A point $\bar{x} \in X$ is called a fixed point of $f(\cdot)$ if the following holds:*

$$f(\bar{x}) = \bar{x}.$$

The following result is called the *Contraction Mapping Theorem*.

Theorem 1.12. *Let (X, d) be a complete metric space and $f : X \to X$ satisfy the following:*

$$d\big(f(x), f(y)\big) \leqslant \alpha d(x, y), \qquad \forall x, y \in X, \tag{1.3}$$

where $\alpha \in (0, 1)$ is a constant. Then $f(\cdot)$ admits a unique fixed point.

Proof. Pick any $x_0 \in X$, define the sequence

$$x_n = f(x_{n-1}), \qquad n \geqslant 1. \tag{1.4}$$

We have the following estimate:

$$d(x_n, x_{n-1}) \leqslant \alpha d(x_{n-1}, x_{n-2}) \leqslant \cdots \leqslant \alpha^{n-1} d(x_1, x_0), \qquad \forall n \geqslant 1.$$

Consequently,

$$d(x_n, x_{n+\ell}) \leqslant \sum_{k=1}^{\ell} d(x_{n+k}, x_{n+k-1}) \leqslant \sum_{k=1}^{\ell} \alpha^{n+k-1} d(x_1, x_0)$$

$$= \alpha^n \Big(\sum_{k=0}^{\ell} \alpha^k \Big) d(x_1, x_0) = \alpha^n \frac{1 - \alpha^{\ell+1}}{1 - \alpha} d(x_1, x_0)$$

$$\leqslant \frac{\alpha^n}{1 - \alpha} d(x_1, x_0) \to 0, \qquad n \to \infty.$$

This means that $\{x_n\}$ is Cauchy. By the completeness of (X, d), one has $x_n \to \bar{x}$ for some $\bar{x} \in X$. Then $f(\bar{x}) = \bar{x}$. Now, if there exists another $\hat{x} \in X$ such that $f(\hat{x}) = \hat{x}$, then

$$d(\bar{x}, \hat{x}) = d(f(\bar{x}), f(\hat{x})) \leqslant \alpha d(\bar{x}, \hat{x}).$$

Since $\alpha < 1$, we must have $\bar{x} = \hat{x}$, proving the uniqueness. $\qquad \square$

In the above, (1.4) is called a *Picard iteration*.

Let us consider two functions $f : X \to Y$ and $g : X \to Z$. We can define their *direct sum* $f \oplus g : X \to Y \times Z$ by the following:

$$(f \oplus g)(x) = (f(x), g(x)), \qquad \forall x \in X.$$

The following can be proved easily.

Proposition 1.13. *Let* (X, d_X), (Y, d_Y), *and* (Z, d_Z) *be metric spaces. Let* $f : X \to Y$ *and* $g : X \to Z$.

(i) *$f(\cdot)$ and $g(\cdot)$ are both continuous at $x_0 \in X$ (resp. on X) if and only if $f \oplus g$ is continuous at x_0 (resp. on X).*

(ii) *In the case that $Y = Z = \mathbb{R}$ with the standard metric, if $f(\cdot)$ and $g(\cdot)$ are continuous at x_0 (resp. on X), the functions $f(\cdot) \pm g(\cdot)$, $f(\cdot)g(\cdot)$, $f(\cdot)/g(\cdot)$, $f(\cdot) \vee g(\cdot) \equiv \max\{f(\cdot), g(\cdot)\}$, and $f(\cdot) \wedge g(\cdot) \equiv \min\{f(\cdot), g(\cdot)\}$ are continuous at x_0 (resp. on X). For $f(\cdot)/g(\cdot)$, the continuity holds in its domain.*

Exercises

1. Prove Proposition 1.4.

2. Let (X, d_X) and (Y, d_Y) be two metric spaces.

(i) Suppose (X, d_X) is discrete. Then any function $f : X \to Y$ must be continuous.

(ii) Suppose $(X, d_X) = (\mathbb{R}^m, \rho_2)$ and (Y, d_Y) are discrete. Then $f : X \to Y$ is continuous if and only if f is a constant function.

3. Let $f : \mathbb{R}^m \to \mathbb{R}^\ell$, with $f(x) = (f_1(x), \cdots, f_\ell(x))$, for all $x \in \mathbb{R}^m$. Then $f(\cdot)$ is continuous if and only if each $f_i(\cdot)$ is continuous, $1 \leqslant i \leqslant \ell$.

4. Let (X, d) be a metric space. Then the metric $d : X \times X \to \mathbb{R}$ is continuous.

5. Let (X, d_X) and (Y, d_Y) be two metric spaces and $f : X \to Y$ such that

$$d_Y(f(x), f(y)) \leqslant L d_X(x, y), \qquad \forall x, y \in X,$$

for some $L > 0$. Then $f(\cdot)$ is continuous.

6. Let $G \subseteq \mathbb{R}^m$ be a domain. Then G is *pathwise connected*, i.e., for any $x, y \in G$, there exists a continuous function $\gamma : [0, 1] \to G$ such that $\gamma(0) = x$ and $\gamma(1) = y$.

7. Find a counterexample showing that a connected set in $\mathbb{R}^m$ might not be pathwise connected.

8. Prove Proposition 1.13.

9. Let $f_\lambda : \mathbb{R} \to \mathbb{R}$ ($\lambda \in \Lambda$) be a family of continuous functions. Let

$$F(x) = \sup_{\lambda \in \Lambda} f_\lambda(x), \qquad x \in \mathbb{R}.$$

Show that $F(\cdot)$ is lower semi-continuous.

2 Properties of Continuous Functions

In this section, we will present several important properties of continuous functions.

2.1 Compactness preserving and uniform continuity

The following result is concerned with the compactness preserving property of continuous functions.

Theorem 2.1. *Let (X, d_X) and (Y, d_Y) be metric spaces. Let $f : X \to Y$ be continuous.*

(i) *Let $K \subseteq X$ be compact. Then*

$$f(K) = \{f(x) \mid x \in K\}$$

is compact in Y.

(ii) *Let $Y = \mathbb{R}$, and (X, d_X) be compact. Then $f(\cdot)$ is bounded and it attains its maximum and minimum.*

Proof. (i) Let $\{y_n\}_{n \geqslant 1}$ be any sequence from $f(K)$. Then there is a sequence $\{x_n\}_{n \geqslant 1} \subseteq K$ such that

$$f(x_n) = y_n, \qquad n \geqslant 1.$$

Since K is compact, there exists a subsequence $\{x_{n_k}\}_{k \geqslant 1}$ such that $x_{n_k} \to \bar{x} \in K$. Then by the continuity of $f(\cdot)$, we have

$$y_{n_k} = f(x_{n_k}) \to f(\bar{x}) \in f(K),$$

which shows the compactness of $f(K)$.

(ii) See the next theorem. $\qquad \square$

The next theorem generalizes (ii) of the above theorem.

Theorem 2.2. *Let (X, d) be a compact metric space and $f : X \to \mathbb{R}$ be lower (resp. upper) semi-continuous. Then $f(\cdot)$ attains its minimum (resp. maximum) on X.*

Proof. Let $x_n \in X$ be a *minimizing sequence*, i.e.,

$$\lim_{n \to \infty} f(x_n) = \inf_{x \in X} f(x).$$

Since X is compact, there exists a subsequence x_{n_k} such that

$$\lim_{k \to \infty} d(x_{n_k}, \bar{x}) = 0,$$

for some $\bar{x} \in X$. Then by the lower semi-continuity of $f(\cdot)$, one has

$$\inf_{x \in X} f(x) \leqslant f(\bar{x}) \leqslant \lim_{n \to \infty} f(x_n) = \inf_{x \in X} f(x).$$

Thus, $\bar{x}$ is a minimum of $f(\cdot)$.

The case of upper semi-continuous function can be proved similarly.

$\square$

We now present an interesting example.

Example 2.3. Let $X = \ell^1$. Let $\bar{B} = \bar{B}(0,1)$ be the closed unit ball in X centered at 0. Define $h : \bar{B} \to \mathbb{R}$ as follows:

$$h(x) = \sum_{k=1}^{\infty} k\big(2|x^k| - 1\big)^+, \qquad \forall x = (x^k) \in \bar{B}, \tag{2.1}$$

where $a^+ = a \vee 0 \equiv \max\{a, 0\}$. Note that for any $x = (x^k) \in \bar{B}$,

$$\sum_{k=1}^{\infty} |x^k| \leqslant 1.$$

Thus, there are at most one non-zero terms in the series (2.1) defining $h(\cdot)$. Now, we show that $h(\cdot)$ is continuous on $\bar{B}$. To see that, we let $\bar{x} = (\bar{x}^k) \in \bar{B}$ be given, and let $x_n = (x_n^k) \in \bar{B}$ such that

$$\rho_1(x_n, \bar{x}) = \sum_{k=1}^{\infty} |x_n^k - \bar{x}^k| \to 0.$$

We may assume that there is a $K \geqslant 1$ such that

$$|\bar{x}^k| < \frac{1}{4}, \qquad \forall k \geqslant K.$$

Then, there exists an $N \geqslant 1$ such that

$$|x_n^k| \leqslant |x_n^k - \bar{x}^k| + |\bar{x}^k| \leqslant \rho_1(x_n, \bar{x}) + |\bar{x}^k| < \frac{1}{2}, \quad n \geqslant N, \ k \geqslant K.$$

Hence,

$$h(x_n) = \sum_{k=1}^{K} k\big(2|x_n^k| - 1\big)^+, \qquad n \geqslant N,$$

with $K \geqslant 1$ independent of $n \geqslant N$. Consequently,

$$|h(x_n) - h(\bar{x})| \leqslant \sum_{k=1}^{K} k\big|\big(2|x_n^k| - 1\big)^+ - \big(2|\bar{x}^k| - 1\big)^+\big|$$

$$\leqslant \sum_{k=1}^{K} 2k|x_n^k - \bar{x}^k| \leqslant 2K \sum_{k=1}^{\infty} |x_n^k - \bar{x}^k| = 2K\rho_1(x_n, \bar{x}) \to 0.$$

This proves the continuity of $h(\cdot)$ at such a point $\bar{x}$. Next, for $x_n = (\delta_{nk})_{k \geqslant 1}$, we have

$$h(x_n) = n \to \infty, \qquad \text{as } n \to \infty.$$

Therefore, $h(\cdot)$ is continuous and unbounded on the bounded and closed set $\bar{B}$. We now define

$$g(x) = \frac{1}{1 + h(x)}, \qquad x \in \bar{B}.$$

Then $g(\cdot)$ is continuous and

$$\inf_{x \in \bar{B}} g(x) = 0, \qquad g(x) > 0, \quad x \in \bar{B}.$$

This means that continuous function $g(\cdot)$ does not attain its minimum over the bounded and closed set $\bar{B}$.

Recall from Chapter 1, Example 4.16 that in ℓ^1, the closed unit ball $\bar{B}(0,1)$ is not compact. Hence, the above example tells us that if the domain space X is not compact, a continuous function defined on a bounded and closed set might not be bounded and might not achieve its minimum and maximum.

Next, let us look at another example which will lead us to another issue.

Example 2.4. (i) Prove the continuity of $f(x) = x^2$ on $[0,1]$.

Let $x_0 \in [0,1]$. One has

$$|x^2 - x_0^2| = |x - x_0||x + x_0| \leqslant 2|x - x_0|.$$

Hence, for any $\varepsilon > 0$, by taking $\delta = \frac{\varepsilon}{2} > 0$, one has

$$|f(x) - f(x_0)| = |x^2 - x_0^2| \leqslant 2|x - x_0| < 2\delta = \varepsilon, \quad \forall |x - x_0| < \delta.$$

We note that $\delta = \frac{\varepsilon}{2}$ is independent of x_0.

(ii) Prove the continuity of $g(x) = \frac{1}{x}$ on $(0,1]$.

Let $x_0 \in (0,1]$. For $|x-x_0| < \frac{x_0}{2}$, we have $x > \frac{x_0}{2}$. Now, for $|x-x_0| < \frac{x_0}{2}$,

$$|g(x) - g(x_0)| = \left| \frac{1}{x} - \frac{1}{x_0} \right| = \frac{|x - x_0|}{x x_0} \leqslant \frac{2}{x_0^2}|x - x_0|.$$

Thus, for any $\varepsilon > 0$, by taking $\delta = \min\left\{ \frac{x_0}{2}, \frac{\varepsilon x_0^2}{2} \right\}$, we have

$$|g(x) - g(x_0)| = \frac{|x - x_0|}{x x_0} \leqslant \frac{2}{x_0^2}|x - x_0| < \varepsilon, \quad \forall |x - x_0| < \delta.$$

This proves the continuity of $g(\cdot)$ on $(0,1]$. In this proof, δ not only depends on $\varepsilon > 0$, but also depends on x_0. We claim that for this function, one cannot find a $\delta > 0$ that is independent of x_0. In fact, if there existed a

$\delta = \delta(\varepsilon)$ independent of x_0, then, say, for $\varepsilon = 1$, we would have a $\delta > 0$ such that

$$\left| g(x) - g\left(\frac{1}{n}\right) \right| = \left| \frac{1}{x} - \frac{1}{1/n} \right| = \frac{|nx - 1|}{x} < 1, \quad \forall \left| x - \frac{1}{n} \right| < \delta.$$

But this is impossible, since we may take $x = \frac{2}{n-1}$ with $n \geqslant 1$ large enough so that

$$0 < \frac{2}{n-1} - \frac{1}{n} = \frac{n+1}{n(n-1)} < \delta.$$

Then for such an x,

$$1 > \frac{|nx - 1|}{x} = \frac{\left| \frac{2n}{n-1} - 1 \right|}{\frac{2}{n-1}} = \frac{n+1}{2}, \qquad \forall n \geqslant 1,$$

which is a contradiction.

The above examples show that sometimes, even for one variable functions, continuity could be quite different. The following notion reveals such a difference.

Definition 2.5. Let (X, d_X) and (Y, d_Y) be metric spaces. A map $f : X \to Y$ is said to be *uniformly continuous* on X if for any $\varepsilon > 0$, there exists a $\delta = \delta(\varepsilon) > 0$, only depends on ε, such that

$$d_Y\big(f(x), f(x')\big) < \varepsilon, \qquad \forall x, x' \in X, \ d_X(x, x') < \delta.$$

The following result gives uniform continuity of continuous functions in metric spaces.

Theorem 2.6. Let (X, d_X) and (Y, d_Y) be metric spaces, with (X, d_X) being compact. Let $f : X \to Y$ be continuous. Then $f(\cdot)$ is uniformly continuous.

Proof. Fix any $\varepsilon > 0$. For each $x_0 \in X$, by the continuity of $f(\cdot)$, there exists a $\delta(x_0) > 0$ such that

$$d_Y\big(f(x), f(x_0)\big) < \frac{\varepsilon}{2}, \qquad x \in B\big(x_0, \delta(x_0)\big).$$

Then for any $x, x' \in B\big(x_0, \delta(x_0)\big)$, we have

$$d_Y\big(f(x), f(x')\big) \leqslant d_Y\big(f(x), f(x_0)\big) + d_Y\big(f(x_0), f(x')\big) < \varepsilon.$$

Clearly,

$$X \subseteq \bigcup_{x_0 \in X} B\big(x_0, \delta(x_0)/2\big).$$

By compactness of X, we can find $x_1, \cdots, x_n \in X$ such that

$$X \subseteq \bigcup_{i=1}^{n} B\big(x_i, \delta(x_i)/2\big).$$

Take

$$\delta = \delta(\varepsilon) = \min_{1 \leqslant i \leqslant n} \delta(x_i)/2 > 0.$$

Now, let $x, x' \in X$ with $d_X(x, x') < \delta$. There must be some $1 \leqslant i \leqslant n$ such that

$$x \in B(x_i, \delta(x_i)/2).$$

Then

$$d_X(x', x_i) \leqslant d_X(x', x) + d_X(x, x_i) < \delta + \frac{\delta(x_i)}{2} \leqslant \delta(x_i).$$

Hence, $x, x' \in B\big(x_i, \delta(x_i)\big)$ which implies

$$d_Y\big(f(x), f(x')\big) < \varepsilon.$$

This proves the uniform continuity of $f(\cdot)$. $\qquad\square$

Let us provide a different proof of the above result.

Proof. Suppose $f(\cdot)$ is not uniformly continuous. Then there exists an $\varepsilon_0 > 0$ such that for any $n \geqslant 1$, there are two points $x_n, y_n \in X$ satisfying

$$d_Y\big(f(x_n), f(y_n)\big) \geqslant \varepsilon_0, \quad d_X(x_n, y_n) < \frac{1}{n}.$$

Since X is compact, there exists a subsequence $\{x_{n_k}\}_{k \geqslant 1}$ of $\{x_n\}_{n \geqslant 1}$ such that for some $\bar{x}$,

$$\lim_{k \to \infty} d_X(x_{n_k}, \bar{x}) = 0.$$

Consequently,

$$d_X(y_{n_k}, \bar{x}) \leqslant d_X(x_{n_k}, y_{n_k}) + d_X(x_{n_k}, \bar{x}) \leqslant \frac{1}{n_k} + d_X(x_{n_k}, \bar{x}) \to 0.$$

Then

$$0 < \varepsilon_0 \leqslant \varlimsup_{k \to \infty} d_Y\big(f(x_{n_k}), f(y_{n_k})\big)$$

$$\leqslant \lim_{k \to \infty} \Big(d_Y\big(f(x_{n_k}), f(\bar{x})\big) + d_Y\big(f(\bar{x}), f(y_{n_k})\big)\Big) = 0,$$

a contradiction. $\qquad\square$

It is not hard to see that in Example 2.3, the continuous function $h(\cdot)$ defined on the closed unit ball is not uniformly continuous. Hence, the compactness condition of X cannot be replaced by boundedness and completeness in the above theorem.

2.2 *Connectedness preserving*

We recall the definition of connected/disconnected sets (Definition 3.1 (x), and its equivalent condition in Proposition 3.9 of Chapter 1). We present the following connectedness preserving property of continuous functions.

Theorem 2.7. *Let (X, d_X) and (Y, d_Y) be two metric spaces, and $f : X \to Y$ be continuous. Let $E \subseteq X$ be connected. Then $f(E)$ is also connected.*

Proof. Suppose $f(E)$ is disconnected. Then by Proposition 3.9 of Chapter 1, there are disjoint non-empty open sets U and V such that

$$f(E) \subseteq U \cup V, \qquad f(E) \cap U \neq \varnothing, \qquad f(E) \cap V \neq \varnothing.$$

Since $f(\cdot)$ is continuous, $f^{-1}(U)$ and $f^{-1}(V)$ are open in X. Further,

$$E \subseteq f^{-1}(U \cup V) = f^{-1}(U) \cup f^{-1}(V).$$

Moreover, $f^{-1}(U)$ and $f^{-1}(V)$ are disjoint, and

$$E \cap f^{-1}(U) \neq \varnothing, \qquad E \cap f^{-1}(V) \neq \varnothing.$$

This means that E is disconnected, a contradiction. $\qquad\qquad\square$

The above result has a well-known corollary which is presented here.

Corollary 2.8. (Intermediate Value Theorem) *Let (X, d_X) be a metric space and $f : X \to \mathbb{R}$ be continuous. Let $E \subseteq X$ be connected, and $a, b \in E$ with*

$$f(a) < f(b).$$

Then for any $y \in (f(a), f(b))$, there exists a $c \in E$ such that

$$f(c) = y.$$

Proof. To prove our conclusion, we first claim that any set $G \subseteq \mathbb{R}$ is connected if and only if it is an interval. In fact, for any $x, y \in G$, we claim that $[x, y] \subseteq G$. By contradiction, if there exists a $z \in (x, y) \setminus G$, then

$$[(-\infty, z) \cap G] \cup [(z, \infty) \cap G] = G.$$

We see that $(-\infty, z) \cap G$ and $(z, \infty) \cap G$ are two disjoint non-empty open sets in G, which means that G is disconnected, a contradiction. Now, let

$$a = \inf G \geqslant -\infty, \quad b = \sup G \leqslant +\infty.$$

Then the above result implies that $(a, b) \subseteq G$. We then have the following situations,

$$G = \begin{cases} (a, b), & a, b \notin G, \\ [a, b), & a \in G, \ b \notin G, \\ (a, b], & a \notin G, \ b \in G, \\ [a, b], & a, b \in G. \end{cases}$$

Thus, G is an interval. Conversely, of course, if G is an interval, then, it is clearly connected.

Now, since E is connected, by Theorem 2.7, $f(E)$ is connected in $\mathbb{R}$, which must be an interval. Hence, $(f(a), f(b)) \subseteq f(E)$. Our conclusion then follows. $\qquad \square$

Exercises

1. Let (X, d_X) and (Y, d_Y) be two metric spaces. Assume that (X, d_X) is compact, and $f : X \to Y$ is bijective and continuous. Then $f^{-1} : Y \to X$ is also continuous. What if (X, d_X) is not compact? Why?

2. Let (X, d) be a compact metric space and $f : X \to X$ be a continuous function such that

$$d(f(x), f(y)) < d(x, y), \qquad \forall x, y \in X, \ x \neq y.$$

Then f has a unique fixed point.

3. Let (X, d) be a compact metric space and $f, g : X \to \mathbb{R}$ be continuous functions such that

$$f(x) \neq g(x), \qquad \forall x \in X.$$

Then there exists an $\varepsilon > 0$ such that

$$|f(x) - g(x)| \geqslant \varepsilon, \qquad \forall x \in X.$$

4. Show that $\overline{\lim}_{x \to 0} \left(\sin \dfrac{1}{x} \right) = 1$ and $\underline{\lim}_{x \to 0} \left(\sin \dfrac{1}{x} \right) = -1$.

5. Find $\lim_{x \to 0} \left[(\sin x) \left(\sin \dfrac{1}{x} \right) \right]$.

6. Let $f : \mathbb{R} \to \mathbb{R}$ and $A_\lambda \subseteq \mathbb{R}$ ($\lambda \in \Lambda$) be a family of sets. Then

$$f^{-1} \left(\bigcup_{\lambda \in \Lambda} A_\lambda \right) = \bigcup_{\lambda \in \Lambda} f^{-1}(A_\lambda), \quad f^{-1} \left(\bigcap_{\lambda \in \Lambda} A_\lambda \right) = \bigcap_{\lambda \in \Lambda} f^{-1}(A_\lambda).$$

7. Suppose $f : [a, b] \to \mathbb{R}$ is continuous such that for any $\xi \in \mathbb{Q} \cap [a, b]$, $f(\xi) = 0$. Then $f(x) = 0$ for all $x \in [a, b]$.

8. Let $f : [a, b] \to \mathbb{R}$ be continuous. Let $a < x_1 < x_2 < \cdots < x_n < b$. Then there exists a $\xi \in [a, b]$ such that

$$f(\xi) = \frac{f(x_1) + f(x_2) + \cdots + f(x_n)}{n}.$$

9. Let $f : [a, b] \to \mathbb{R}$ be continuous such that $f(x) \neq 0$ for all $x \in [a, b]$, and there exists a $\xi \in [a, b]$ such that $f(\xi) > 0$. Then $f(x) > 0$ for all $x \in [a, b]$.

10. Suppose $f : [a, b] \to \mathbb{R}$ is an increasing function and for any $y \in [f(a), f(b)]$, there exists a $\xi \in [a, b]$ such that $f(\xi) = y$. Then $f(\cdot)$ is continuous on $[a, b]$. What happens if $f(\cdot)$ is not monotone? Why?

11. Let $f : (a, b) \to \mathbb{R}$ be lower semi-continuous such that

$$\lim_{x \to a+0} f(x) = \lim_{x \to b-0} f(x) = \infty.$$

Then there exists a $\xi \in (a, b)$ such that

$$f(\xi) \leqslant f(x), \qquad \forall x \in (a, b).$$

12. Let $S \subseteq \mathbb{R}$ (not necessarily bounded) and $f, g : S \to \mathbb{R}$ be uniformly continuous. Then $f \pm g$ are uniformly continuous on S. Is $f \cdot g$ also uniformly continuous? Why?

13. Let S be a bounded set in $\mathbb{R}$ (not necessarily closed) and $f : S \to \mathbb{R}$ be a uniformly continuous function. Then f is bounded on S. What if S is not bounded, why?

14. Let $f : (a, b) \to \mathbb{R}$ be continuous. Then $f(\cdot)$ is uniformly continuous on (a, b) if and only if both $\lim\limits_{x \to a^+} f(x)$ and $\lim\limits_{x \to b^-} f(x)$ exist and finite.

15. Prove or disprove that the following functions are uniformly continuous in their domains:

(i) $f(x) = \sin \frac{1}{x}$ on $(0, \pi]$;

(ii) $f(x) = \sin x$ on $(-\infty, \infty)$;

(iii) $f(x) = x^2$ on $(-\infty, \infty)$.

16. Suppose $f : \mathbb{R} \to \mathbb{R}$ is uniformly continuous. Then there exists a constant $K > 0$ such that

$$|f(x)| \leqslant K(1 + |x|), \qquad \forall x \in \mathbb{R}.$$

Chapter 3

Differentiation

1 Derivatives

Throughout of this chapter, $G \subseteq \mathbb{R}^m$ will be a *domain* (a connected and open set). Let $f : G \to \mathbb{R}$ and $x_0 \in G$ be given. We would like to look at the behavior of $f(\cdot)$ near the point x_0 (by which we mean in some small neighborhood of x_0) beyond the limit $\lim_{x \to x_0} f(x)$ and the continuity. To this end, we introduce the following definition.

Definition 1.1. Let $f : G \to \mathbb{R}$.

(i) Let $x_0 \in G$, and let $u \in S_1^m(0) \triangleq \{u \in \mathbb{R}^m \mid \|u\| = 1\}$. Suppose

$$f'(x_0; u) \equiv \lim_{\delta \downarrow 0} \frac{f(x_0 + \delta u) - f(x_0)}{\delta} \tag{1.1}$$

exists. Then we say that $f(\cdot)$ is *directionally differentiable* at x_0 in the direction u, and $f'(x_0; u)$ is called a *directional derivative* of $f(\cdot)$ at x_0 in the direction of u. If $f'(x_0; u)$ exists for all $u \in S_1^m(0)$, we say that $f(\cdot)$ is *Gâteaux differentiable* at x_0, and call the map $u \mapsto f'(x_0; u)$ the *Gâteaux derivative* of $f(\cdot)$ at x_0.

(ii) If for $x_0 \in G$, $f'(x_0; u)$ exists for every $u \in S_1^m(0)$, and there exists a row vector, denoted by $(\nabla f)(x_0)$, such that

$$f'(x_0; u) = (\nabla f)(x_0)u, \qquad \forall u \in S_1^m(0), \tag{1.2}$$

then $(\nabla f)(x_0)$ is called the *gradient* of $f(\cdot)$ at x_0. If there exists a function, denoted by $\nabla f : G \to \mathbb{R}^{1 \times m}$ such that (1.2) holds for all $x_0 \in G$, then we call ∇f the *gradient field* of $f(\cdot)$.

(iii) If for $x_0 \in G$, there exists a row vector, denoted by $f_x(x_0) \in \mathbb{R}^{1 \times m}$ such that

$$\lim_{\|y\| \to 0} \frac{|f(x_0 + y) - f(x_0) - f_x(x_0)y|}{\|y\|} = 0, \tag{1.3}$$

61

then we say that $f(\cdot)$ is *Fréchet differentiable* at x_0, and $f_x(x_0)$ is called the *Fréchet derivative* of $f(\cdot)$ at x_0. In the case $m = 1$, we usually denote $f_x(x_0)$ by $f'(x_0)$.

From the above definition, we see that if $f(\cdot)$ is directionally differentiable at x_0 in the direction $u \in S_1^m(0)$, then

$$f'(x_0; \lambda u) \triangleq \lim_{\delta \to 0} \frac{f(x_0 + \delta \lambda u) - f(x_0)}{\delta} = \lambda f'(x_0; u), \quad \forall \lambda \geqslant 0. \tag{1.4}$$

Consequently, if $f(\cdot)$ is Gâteaux differentiable at x_0, then the Gâteaux derivative $u \mapsto f'(x_0; u)$ can be extended to a map from $\mathbb{R}^m$ to $\mathbb{R}$, which is positively homogeneous. However, this map might not be linear. Whereas, if $f(\cdot)$ admits a gradient $(\nabla f)(x_0)$, then the map $u \mapsto f'(x_0; u) \equiv (\nabla f)(x_0)u$ is linear. The following example shows that Gâteaux differentiability does not necessarily imply the existence of a gradient.

Example 1.2. Let $m = 2$ and

$$f(x, y) = \begin{cases} \dfrac{xy}{\sqrt{x^2 + y^2}}, & (x, y) \neq (0, 0), \\ 0, & (x, y) = (0, 0). \end{cases}$$

For any $(\alpha, \beta) \in \mathbb{R}^2$, we have

$$f'((0,0); (\alpha, \beta)) = \lim_{\delta \downarrow 0} \frac{f(\delta \alpha, \delta \beta) - f(0,0)}{\delta} = \frac{\alpha \beta}{\sqrt{\alpha^2 + \beta^2}},$$

which is positively homogeneous in (α, β), but it is not linear. Hence, $(\nabla f)(0,0)$ does not exists, although $f(\cdot)$ is Gâteaux differentiable at $(0,0)$.

As a matter of fact, even in the case $m = 1$, Gâteaux differentiability does not implies the existence of a gradient. An easy counterexample is $f(x) = |x|$ at $x = 0$. We have

$$f'(0; \lambda) = |\lambda|, \quad \lambda \in \mathbb{R},$$

which is not linear in λ. Thus, $(\nabla f)(0)$ fails to exist.

Next, we have the following result.

Proposition 1.3. *Let* $f : G \to \mathbb{R}$ *and* $x_0 \in G$. *If* $f(\cdot)$ *is Fréchet differentiable at* x_0, *then* $f(\cdot)$ *admits a gradient* $(\nabla f)(x_0)$. *Moreover,*

$$(\nabla f)(x_0) = f_x(x_0). \tag{1.5}$$

Consequently, for any $u \in S_1^m(0)$, *the directional derivative of* $f(\cdot)$ *in the direction* u *at* x_0 *is given by*

$$f'(x_0; u) = f_x(x_0)u. \tag{1.6}$$

Proof. For any $u \in S_1^m(0)$, by (1.3), we have

$$\lim_{\delta \downarrow 0} \frac{|f(x_0 + \delta u) - f(x_0) - \delta f_x(x_0)u|}{\delta} = 0, \qquad (1.7)$$

which implies that

$$\lim_{\delta \downarrow 0} \frac{f(x_0 + \delta u) - f(x_0)}{\delta} = f_x(x_0)u. \qquad (1.8)$$

Thus, by definition, we see that the gradient $(\nabla f)(x_0)$ exists at x_0 and (1.5) holds. $\square$

From Definition 1.1, we see that for $m = 1$, the existence of a gradient and the Fréchet differentiability are equivalent. However, we claim that the existence of a gradient does not necessarily imply the Fréchet differentiability, for $m \geqslant 2$. To show this, let us first present the following result.

Proposition 1.4. *If* $f : G \to \mathbb{R}$ *is Fréchet differentiable at* x_0, *then it is continuous at* x_0.

Proof. By definition, we have that

$$\lim_{x \to x_0} \frac{|f(x) - f(x_0) - f_x(x_0)(x - x_0)|}{\|x - x_0\|} \equiv \lim_{x \to x_0} \frac{|R_1(x, x_0)|}{\|x - x_0\|} = 0, \qquad (1.9)$$

where

$$R_1(x, x_0) = f(x) - f(x_0) - f_x(x_0)(x - x_0). \qquad (1.10)$$

Thus,

$$\lim_{x \to x_0} |f(x) - f(x_0)| \leqslant \lim_{x \to x_0} \left\{ |R_1(x, x_0)| + |f_x(x_0)(x - x_0)| \right\}$$
$$\leqslant \lim_{x \to x_0} \left\{ \frac{|R_1(x, x_0)|}{\|x - x_0\|} + \|f_x(x_0)\| \right\} \|x - x_0\| = 0, \qquad (1.11)$$

proving the continuity of $f(\cdot)$ at x_0. $\square$

Note that (1.10) can also be written as

$$f(x) = f(x_0) + f_x(x_0) \cdot (x - x_0) + R_1(x, x_0), \qquad (1.12)$$

with the *remainder term* $R_1(x, x_0)$ given by (1.10). We refer to (1.12) as a *first order Taylor expansion*. We will come back to this a little later.

The above result will help us to see that having gradient does not necessarily imply the Fréchet differentiability. Here is an example.

Example 1.5. Let $m = 2$ and

$$f(x, y) = \begin{cases} \dfrac{x^3 y}{x^6 + y^2}, & (x, y) \neq (0, 0), \\ 0, & (x, y) = (0, 0). \end{cases} \qquad (1.13)$$

Then for any $(\alpha, \beta) \in \mathbb{R}^2$ with $\beta \neq 0$, we have

$$\lim_{\delta \downarrow 0} \frac{f(\delta\alpha, \delta\beta) - f(0,0)}{\delta} = \lim_{\delta \downarrow 0} \frac{\delta^2 \alpha^3 \beta}{\delta^4 \alpha^6 + \beta^2} = 0,$$

and for any $(\alpha, 0) \in \mathbb{R}^2$,

$$\lim_{\delta \downarrow 0} \frac{f(\delta\alpha, 0) - f(0,0)}{\delta} = 0.$$

Thus, we have

$$(\nabla f)(0,0) = (0,0).$$

On the other hand, by taking a sequence $\{(x_n, y_n)\}$ approaching to $(0,0)$ along the curve $y = ax^3$ for some $a \in \mathbb{R}$, we see that

$$\lim_{n \to \infty} f(x_n, y_n) = \frac{a}{1 + a^2}.$$

Hence, for different choices of $a \in \mathbb{R}$, the above limit is different. This means that $f(x, y)$ does not have a limit at $(0,0)$, which implies that $f(\cdot, \cdot)$ is not continuous at $(0,0)$. By Proposition 1.4, $f(\cdot, \cdot)$ must not be Fréchet differentiable at $(0,0)$.

Note that in Definition 1.1, various differentiabilities were introduced without using the coordinates of the space. Now, in the case that one uses the coordinates, some more interesting features will be revealed.

Let $e_i \in \mathbb{R}^m$ be the vector whose i-th component is 1 and all others are 0. We call $\{e_1, \cdots, e_m\}$ the *standard orthonormal basis* for $\mathbb{R}^m$. If $f(\cdot)$ admits a directional derivative at x_0 in the directions of $\pm e_i$, and

$$f'(x_0; -e_i) = -f'(x_0; e_i), \tag{1.14}$$

then we denote

$$f_{x^i}(x_0) = f'(x_0; e_i) \equiv \lim_{\delta \downarrow 0} \frac{f(x_0 + \delta e_i) - f(x_0)}{\delta}, \tag{1.15}$$

and call it the *(first order) partial derivative* of $f(\cdot)$ at x_0 with respect to x^i. Note that when (1.14) holds, we have

$$\lim_{\delta \to 0^+} \frac{f(x_0 + \delta e_i) - f(x_0)}{\delta} = \lim_{\delta \to 0^-} \frac{f(x_0 + \delta e_i) - f(x_0)}{\delta}. \tag{1.16}$$

Thus,

$$f_{x^i}(x_0) = \lim_{\delta \to 0} \frac{f(x_0 + \delta e_i) - f(x_0)}{\delta}. \tag{1.17}$$

Further, if $m = 1$, we simply denote

$$f'(x_0) = f_{x^1}(x_0), \tag{1.18}$$

and call it the *derivative* of $f(\cdot)$ at x_0.

The following result gives a representation of gradient as well as directional derivatives in terms of partial derivatives.

Proposition 1.6. *Let* $f : G \to \mathbb{R}$, $x_0 \in G$, *and* $f(\cdot)$ *admit gradient* $(\nabla f)(x_0)$ *at* x_0. *Then* $f(\cdot)$ *has partial derivatives* $f_{x^i}(x_0)$, $1 \leqslant i \leqslant m$. *In this case, under the standard orthonormal basis of* $\mathbb{R}^m$, *the gradient* $(\nabla f)(x_0)$ *admits the following representation:*

$$(\nabla f)(x_0) = (f_{x^1}(x_0), \cdots, f_{x^m}(x_0)). \tag{1.19}$$

Consequently, for any direction $u \equiv (u^1, \cdots, u^m)^\top \in S_1^m(0)$,

$$f'(x_0; u) = \sum_{i=1}^m u^i f_{x^i}(x_0). \tag{1.20}$$

Proof. Taking $u = \pm e_i$ in (1.1), we see that $f_{x^i}(x_0)$ exists for each $1 \leqslant i \leqslant m$. Next, for any $u \in S_1^m(0)$, we have

$$u = \sum_{i=1}^m u^i e_i, \tag{1.21}$$

with $u^i = u \cdot e_i$ $(1 \leqslant i \leqslant m)$. Then

$$(\nabla f)(x_0)u = \sum_{i=1}^m u^i (\nabla f)(x_0)e_i = \sum_{i=1}^m u^i f_{x^i}(x_0)$$

$$= (f_{x^1}(x_0), \cdots, f_{x^m}(x_0)) \begin{pmatrix} u^1 \\ \vdots \\ u^m \end{pmatrix}. \tag{1.22}$$

This implies the representation (1.19) of $(\nabla f)(x_0)$ under the standard orthonormal basis of $\mathbb{R}^m$. Then (1.20) follows from (1.2) and (1.19). $\qquad \square$

Let us go back to look at Example 1.2. It is clear that

$$f_x(0,0) = f_y(0,0) = 0,$$

but $\nabla f(0,0)$ does not exist. Thus, one could have existence of all first order partial derivatives without the existence of the gradient.

The following result collects some basic properties of the gradients.

Proposition 1.7. *Let* $f, g : G \to \mathbb{R}$ *admit gradients at* $x_0 \in G$. *Then*

(i) *for any* $\lambda, \mu \in \mathbb{R}$, $\lambda f(\cdot) + \mu g(\cdot)$ *admits a gradient at* x_0 *and*

$$[\nabla(\lambda f + \mu g)](x_0) = \lambda(\nabla f)(x_0) + \mu(\nabla g)(x_0). \qquad (1.23)$$

(ii) *the product* $f(\cdot)g(\cdot)$ *admits a gradient at* x_0 *and*

$$[\nabla(fg)](x_0) = g(x_0)(\nabla f)(x_0) + f(x_0)(\nabla g)(x_0). \qquad (1.24)$$

(iii) *the reciprocal* $f(\cdot)^{-1}$ *admits a gradient at* x_0, *provided* $f(x_0) \neq 0$, *and in this case,*

$$[\nabla(f^{-1})](x_0) = -\frac{(\nabla f)(x_0)}{f(x_0)^2}. \qquad (1.25)$$

The proof is straightforward. We also have similar result with the gradient replaced by any first order partial derivative. We omit the details here.

Next, we look at the vector-valued function case. Similar to the scalar case, we introduce the following.

Definition 1.8. Let $f : G \to \mathbb{R}^\ell$.

(i) Let $x_0 \in G$, and let $u \in S_1^m(0)$. Suppose

$$f'(x_0; u) \equiv \lim_{\delta \downarrow 0} \frac{f(x_0 + \delta u) - f(x_0)}{\delta} \qquad (1.26)$$

exists. Then we say that $f(\cdot)$ admits a *directional derivative* $f'(x_0; u)$ at x_0 in the direction of u. If $f'(x_0; u)$ exists for all $u \in S_1^m(0)$, we say that $f(\cdot)$ is *Gâteaux differentiable*.

(ii) If for $x_0 \in G$, there exists an $(\ell \times m)$ matrix, denoted by $(\nabla f)(x_0) \in \mathbb{R}^{\ell \times m}$ such that

$$f'(x_0; u) = (\nabla f)(x_0)u, \qquad \forall u \in S_1^m(0), \qquad (1.27)$$

then $(\nabla f)(x_0)$ is called the *gradient* of $f(\cdot)$ at x_0. If there exists an $(\ell \times m)$ matrix-valued function, denoted by $\nabla f : G \to \mathbb{R}^m$ such that (1.2) holds for all $x_0 \in G$, then we call ∇f the *gradient field* of $f(\cdot)$.

(iii) If for $x_0 \in G$, there exists an $(\ell \times n)$ matrix, denoted by $f_x(x_0) \in \mathbb{R}^{\ell \times m}$ such that

$$\lim_{\|y\| \to 0} \frac{\|f(x_0 + y) - f(x_0) - f_x(x_0)y\|}{\|y\|} = 0, \qquad (1.28)$$

then we say that $f(\cdot)$ is *Fréchet differentiable* at x_0. Further, if $f(\cdot)$ is Fréchet differentiable in a neighborhood of $x_0 \in G$ and $f_x(\cdot)$ is continuous in that neighborhood, we say that $f(\cdot)$ is *continuously Fréchet differentiable* near x_0, or simply $f(\cdot)$ is C^1 near x_0.

The following result links the (directional, Gâteaux, and Fréchet) differentiability of a vector-valued function with its scalar function components.

Proposition 1.9. *Let $f \equiv (f^1, \cdots, f^\ell) : G \to \mathbb{R}^\ell$ and $x_0 \in G$. Then*

(i) $f(\cdot)$ *is directionally differentiable at x_0 in the direction $u \in S_1^m(0)$ if and only if so is each $f^i(\cdot)$, and in this case,*

$$f'(x_0; u) = \left((f^1)'(x_0; u), \cdots, (f^\ell)'(x_0; u) \right)^\top. \tag{1.29}$$

(ii) $f(\cdot)$ *is Gâteaux differentiable at x_0 if and only if so is each $f^i(\cdot)$.*

(iii) $f(\cdot)$ *admits a gradient at x_0 if and only if so is each $f^i(\cdot)$, and in this case,*

$$(\nabla f)(x_0) = \begin{pmatrix} (\nabla f^1)(x_0) \\ \vdots \\ (\nabla f^\ell)(x_0) \end{pmatrix} \in \mathbb{R}^{\ell \times m}. \tag{1.30}$$

(iv) $f(\cdot)$ *is Fréchet differentiable at x_0 if and only if so is each $f^i(\cdot)$, and in this case,*

$$f_x(x_0) = \begin{pmatrix} f_x^1(x_0) \\ \vdots \\ f_x^\ell(x_0) \end{pmatrix} \in \mathbb{R}^{\ell \times m}. \tag{1.31}$$

Note that under standard orthonormal bases of $\mathbb{R}^m$ and $\mathbb{R}^\ell$, we can write more explicitly that

$$f_x(x_0) = \begin{pmatrix} f_{x^1}^1(x_0) & f_{x^2}^1(x_0) & \cdots & f_{x^m}^1(x_0) \\ f_{x^1}^2(x_0) & f_{x^2}^2(x_0) & \cdots & f_{x^m}^2(x_0) \\ \vdots & \vdots & \cdots & \vdots \\ f_{x^1}^\ell(x_0) & f_{x^2}^\ell(x_0) & \cdots & f_{x^m}^\ell(x_0) \end{pmatrix} \in \mathbb{R}^{\ell \times m}. \tag{1.32}$$

The above is called the *Jacobian matrix* of $f(\cdot)$ with respect to x (at x_0). In the case $\ell = m$, the Jacobian matrix $f_x(x_0) \in \mathbb{R}^{m \times m}$ is a square matrix, and we may define its determinant $\det[f_x(x_0)]$ which is called the *Jacobian (determinant)* of $f(\cdot)$ with respect to x (at x_0). Sometimes, one denotes

$$\det[f_x(x_0)] = \frac{\partial f}{\partial x}(x_0) \equiv \frac{\partial(f^1, \cdots, f^m)}{\partial(x^1, \cdots, x^m)}. \tag{1.33}$$

Next, we look at the derivative of composition of functions.

Theorem 1.10. (Chain Rule) (i) *Let $g : G \to \mathbb{R}^\ell$ be directionally differentiable at $x_0 \in G$ in the direction of $u \in S_1^m(0)$ and $f : \mathbb{R}^\ell \to \mathbb{R}^k$*

be Fréchet differentiable at $g(x_0)$. Then $f \circ g : G \to \mathbb{R}^k$ is directionally differentiable at $x_0 \in G$ in the direction of $u \in S_1^m(0)$, and

$$(f \circ g)'(x_0; u) = f_y(g(x_0))g'(x_0; u). \tag{1.34}$$

(ii) Let $g : G \to \mathbb{R}^\ell$ be Gâteaux differentiable (resp. admitting a gradient, Fréchet differentiable) at $x_0 \in G$ and $f : \mathbb{R}^\ell \to \mathbb{R}^k$ be Fréchet differentiable at $g(x_0)$. Then $f \circ g : G \to \mathbb{R}^k$ is Gâteaux differentiable (resp. admitting a gradient, Fréchet differentiable) at $x_0 \in G$, and

$$(f \circ g)'(x_0; u) = f_y(g(x_0))g'(x_0; u), \qquad \forall u \in \mathbb{R}^m. \tag{1.35}$$

$$(\text{resp. } [\nabla(f \circ g)](x_0) = f_y(g(x_0))(\nabla g)(x_0), \tag{1.36}$$

$$(f \circ g)_x(x_0) = f_y(g(x_0))g_x(x_0)) \tag{1.37}$$

Proof. We only prove (i). By the Fréchet differentiability of $f(\cdot)$ at $g(x_0)$, we know that

$$f(y) = f(g(x_0)) + f_y(g(x_0))(y - g(x_0)) + R_1(y; g(x_0)), \tag{1.38}$$

with

$$\lim_{y \to g(x_0)} \frac{\|R_1(y; g(x_0))\|}{\|y - g(x_0)\|} = 0. \tag{1.39}$$

Now, by taking $y = g(x_0 + \delta u)$, we have

$$\begin{aligned}
&f(g(x_0 + \delta u)) - f(g(x_0)) \\
&= f_y(g(x_0))[g(x_0 + \delta u) - g(x_0)] + R_1(g(x_0 + \delta u); g(x_0)).
\end{aligned} \tag{1.40}$$

Hence,

$$\begin{aligned}
&\lim_{\delta \downarrow 0} \left\| \frac{f(g(x_0 + \delta u)) - f(g(x_0))}{\delta} - f_y(g(x_0))g'(x_0; u) \right\| \\
&= \lim_{\delta \downarrow 0} \left\| f_y(g(x_0)) \left[\frac{g(x_0 + \delta u) - g(x_0)}{\delta} - g'(x_0; u) \right] + \frac{R_1(g(x_0 + \delta u); g(x_0))}{\delta} \right\| \\
&\leqslant \|f_y(g(x_0))\| \lim_{\delta \downarrow 0} \left\| \frac{g(x_0 + \delta u) - g(x_0)}{\delta} - g'(x_0; u) \right\| \\
&\quad + \lim_{\delta \to 0} \frac{\|R_1(g(x_0 + \delta u); g(x_0))\|}{\|g(x_0 + \delta u) - g(x_0)\|} \frac{\|g(x_0 + \delta u) - g(x_0)\|}{\delta} = 0.
\end{aligned}$$

This proves (1.34). $\square$

We note that in the above, $f(\cdot)$ is assumed to be Fréchet differentiable. A natural question is whether we can relax this condition. In other words, naively, we expect the following: If $f(\cdot)$ has a gradient at $g(x_0)$ and $g(\cdot)$

is Fréchet differentiable at x_0, then it seems that the following should be true:

$$[\nabla(f \circ g)](x_0) = (\nabla f)(g(x_0))g_x(x_0). \tag{1.41}$$

Unfortunately, the above is not true in general. Here is a counterexample.

Example 1.11. Let $f(\cdot)$ be defined by (1.13). Then, from Example 1.5, we know that $\nabla f(0,0)$ exists and $f(\cdot)$ is not Fréchet differentiable at $(0,0)$. Now, let

$$g(z) = (z, z^3)^\top, \qquad z \in \mathbb{R}.$$

Then

$$\nabla f(g(0))g_z(0) = \nabla f(0,0) \begin{pmatrix} 1 \\ 0 \end{pmatrix} = 0.$$

Whereas,

$$f(g(z)) = \begin{cases} \dfrac{1}{2}, & z \neq 0, \\ 0, & z = 0, \end{cases}$$

which is not directionally differentiable at $(0,0)$ in any direction. Hence, (1.41) does not hold.

The following is usually referred to as the *Fermat Theorem*.

Theorem 1.12. (Fermat) Suppose $f : G \to \mathbb{R}$ attains a local maximum (or minimum) at point $x_0 \in G$ at which $f(\cdot)$ admits a gradient. Then

$$(\nabla f)(x_0) = 0. \tag{1.42}$$

Proof. Suppose $f(\cdot)$ attains a local maximum at x_0. By definition, this means that there exists an open ball $B(x_0, \sigma) \subseteq G$ such that

$$f(x) \leqslant f(x_0), \qquad \forall x \in B(x_0, \sigma). \tag{1.43}$$

Thus, for any $u \in S_1^m(0)$, we have

$$(\nabla f)(x_0) \cdot u = \lim_{\delta \downarrow 0} \frac{f(x_0 + \delta u) - f(x_0)}{\delta} \leqslant 0. \tag{1.44}$$

Consequently, one must have (1.42).

The case of local minimum can be proved similarly. $\qquad\square$

Any point $x_0 \in G$ satisfying (1.42) is called a *critical point*. Thus, a critical point is a candidate for a local maximum or a local minimum.

Note that in the above, we have only assumed the existence of a gradient $(\nabla f)(x_0)$ of $f(\cdot)$ at x_0 (not the Fréchet differentiability).

Theorem 1.13. (Cauchy Mean Value Theorem) *Let $a, b \in G \subseteq \mathbb{R}^m$, with $a \neq b$, such that*

$$[a, b] \triangleq \{\lambda a + (1 - \lambda)b \mid \lambda \in [0, 1]\} \subseteq G. \tag{1.45}$$

Let $f : [a, b] \to \mathbb{R}^\ell$ and $g : [a, b] \to \mathbb{R}$ be continuously Fréchet differentiable. Then for any $w \in \mathbb{R}^\ell$, there exists a $\xi \in (a, b) \equiv [a, b] \setminus \{a, b\}$ (depending on w) such that

$$g_x(\xi)(b - a)[f(b) - f(a)] \cdot w = [g(b) - g(a)][f_x(\xi)(b - a)] \cdot w. \tag{1.46}$$

Note that if

$$g_x(\xi)(b - a) \neq 0,$$

then (1.46) can be written as

$$[f(b) - f(a)] \cdot w = \frac{f_x(\xi)(b - a)}{g_x(\xi)(b - a)} \{[g(b) - g(a)]w\}. \tag{1.47}$$

There are several useful special cases of the above.

(i) $g(x) = (b - a) \cdot x$. In this case, with $a \neq b$,

$$g_x(x) = (b - a)^\top, \quad \forall x \in \mathbb{R}^m; \qquad g(b) - g(a) = \|b - a\|^2.$$

Thus, (1.46) is equivalent to the following:

$$[f(b) - f(a)] \cdot w = [f_x(\xi)(b - a)] \cdot w. \tag{1.48}$$

From this, by taking $w = \frac{f(b) - f(a)}{\|f(b) - f(a)\|}$ (assuming that $f(b) - f(a) \neq 0$), we have

$$\|f(b) - f(a)\| = [f_x(\xi)(b - a)] \cdot \frac{f(b) - f(a)}{\|f(b) - f(a)\|}, \tag{1.49}$$

which leads to

$$\|f(b) - f(a)\| \leqslant \|f_x(\xi)(b - a)\|. \tag{1.50}$$

(ii) $g(x) = (b - a) \cdot x$ and $\ell = 1$. In this case, by taking $w = 1$, (1.48) becomes

$$f(b) - f(a) = f_x(\xi)(b - a). \tag{1.51}$$

This is called the *Lagrange Mean Value Theorem* for multi-variable functions. A further special case of this is $G = [a, b]$ and $f(a) = f(b)$, which leads to the well-known *Rolle's Theorem*. We also point out that for the case $m = \ell = 1$, since the Fréchet differentiability is equivalent to the existence of a derivative, one has (1.51) under the condition that $f(\cdot)$ admits a derivative at every point on the interval (a, b), without assuming the continuity of the derivative.

(iii) $G = [a, b] \subseteq \mathbb{R}$ and $\ell = 1$. In this case, by taking $w = 1$, we see that (1.46) is equivalent to the following:

$$\frac{f'(\xi)}{g'(\xi)} = \frac{f(b) - f(a)}{g(b) - g(a)}. \tag{1.52}$$

This is also referred to as the *Cauchy Mean Value Theorem*.

Proof of Theorem 1.13. First, if

$$g(b) - g(a) = 0, \tag{1.53}$$

we let

$$h(t) = g(a + t(b - a)) - g(a), \qquad t \in [0, 1]. \tag{1.54}$$

Then $h : [0, 1] \to \mathbb{R}$ is continuous and Fréchet differentiable. Moreover, $h(0) = h(1) = 0$. Thus, by Rolle's Theorem,[1] there exists a $\theta \in (0, 1)$ such that

$$0 = h'(\theta) = g_x(a + \theta(b - a)) \cdot (b - a). \tag{1.55}$$

Hence, with $\xi = a + \theta(b - a) \in (a, b)$, both sides of (1.46) are zero.

Next, let (1.53) fail. For any $w \in \mathbb{R}^\ell$, let

$$h(t) = \Big[f(a + t(b - a)) - f(a) \Big] \cdot w$$
$$- \frac{[f(b) - f(a)] \cdot w}{g(b) - g(a)} [g(a + t(b - a)) - g(a)], \qquad t \in [0, 1]. \tag{1.56}$$

Then $h : [0, 1] \to \mathbb{R}$ is continuous and admits derivatives on $(0, 1)$. Moreover, we again have $h(0) = h(1) = 0$. Thus, by Rolle's Theorem, there exists a $\theta \in (0, 1)$ such that

$$0 = h'(\theta) = [f_x(a + \theta(b - a))(b - a)] \cdot w$$
$$- \frac{[f(b) - f(a)] \cdot w}{g(b) - g(a)} g_x(a + \theta(b - a)) \cdot (b - a). \tag{1.57}$$

Thus, if we let $\xi = a + \theta(b - a) \in (a, b)$, then (1.46) follows. $\qquad \square$

---

[1] We assume that Rolle's Theorem is known to the readers.

Note that by definition, we see that if $f : G \to \mathbb{R}$ is Fréchet differentiable at $x_0 \in G$, then near x_0, $f(x)$ can be approximated by a linear function:

$$f(x) = f(x_0) + f_x(x_0) \cdot (x - x_0) + R(x; x_0) \equiv L(x; x_0) + R(x; x_0),$$

where

$$\begin{cases} L(x; x_0) = f(x_0) + f_x(x_0) \cdot (x - x_0), \\ R(x; x_0) = f(x) - f(x_0) - f_x(x_0) \cdot (x - x_0), \end{cases}$$

with

$$\lim_{\|x - x_0\| \to 0} \frac{|R(x; x_0)|}{\|x - x_0\|} = 0.$$

Whereas, by the Lagrange mean value theorém, we have

$$f(x) = f(x_0) + f_x(x_0 + \theta(x - x_0)) \cdot (x - x_0), \tag{1.58}$$

in which the remainder term does not appear, and as a trade-off, $f_x(x_0)$ is replaced by $f_x(x_0 + \theta(x - x_0))$.

As an application of the above Theorem 1.13, we present the following result.

Proposition 1.14. *Let $f : G \to \mathbb{R}^\ell$ be Fréchet differentiable. Then*

$$f_x(x) = 0, \qquad x \in G, \tag{1.59}$$

if and only if $f(x)$ is a constant.

Proof. Sufficiency is clear. We now prove the necessity. Since G is a domain (connected and open in $\mathbb{R}^\ell$), it must be pathwise connected (see Problem 6 in Chapter 2, Section 1), for any two points $x_0, \bar{x} \in G$, we can find a finite sequence $x_1, \cdots, x_k \in G$ so that $x_k = \bar{x}$ and the segments

$$[x_i, x_{i+1}] \subseteq G, \qquad 0 \leqslant i \leqslant k - 1.$$

By (1.50), we obtain

$$\|f(x_1) - f(x_0)\| \leqslant \|f_x(\xi)(x_1 - x_0)\| = 0.$$

Use the same argument recursively, we obtain that

$$f(\bar{x}) = f(x_0), \qquad \forall \bar{x} \in G,$$

proving our claim. $\qquad \square$

The following is another very interesting application of the Cauchy mean value theorem.

Theorem 1.15. (l'Hôpital's Rule) *Let $f, g : [a, b] \to \mathbb{R}$ be continuous. Let $x_0 \in [a, b]$ such that*

$$f(x_0) = g(x_0) = 0. \tag{1.60}$$

(i) *Suppose $f(\cdot)$ and $g(\cdot)$ have derivatives $f'(x_0)$ and $g'(x_0)$ at x_0 such that $g'(x_0) \neq 0$. Then there exists a $\delta > 0$ such that*

$$g(x) \neq 0, \qquad 0 < |x - x_0| < \delta, \tag{1.61}$$

and

$$\lim_{x \to x_0} \frac{f(x)}{g(x)} = \frac{f'(x_0)}{g'(x_0)}. \tag{1.62}$$

(ii) *Suppose $x_0 = a$ and there exists a $\delta > 0$ such that $f'(x)$ and $g'(x)$ exist for all $x \in (x_0, x_0 + \delta)$ and*

$$g'(x) \neq 0, \qquad x \in (x_0, x_0 + \delta). \tag{1.63}$$

Then

$$\lim_{x \to x_0^+} \frac{f(x)}{g(x)} = \lim_{x \to x_0^+} \frac{f'(x)}{g'(x)}, \tag{1.64}$$

provided the limit on right-hand side exists.

(iii) *Suppose $x_0 = b$ and there exists a $\delta > 0$ such that $f'(x)$ and $g'(x)$ exist for all $x \in (x_0 - \delta, x_0)$ and*

$$g'(x) \neq 0, \qquad x \in (x_0 - \delta, x_0). \tag{1.65}$$

Then

$$\lim_{x \to x_0^-} \frac{f(x)}{g(x)} = \lim_{x \to x_0^-} \frac{f'(x)}{g'(x)}, \tag{1.66}$$

provided the limit on right-hand side exists.

(iv) *Suppose $x_0 \in (a, b)$ and there exists a $\delta > 0$ such that $f'(x)$ and $g'(x)$ exist for all $x \in (x_0 - \delta, x_0 + \delta) \setminus \{x_0\}$ and*

$$g'(x) \neq 0, \qquad x \in (x_0 - \delta, x_0 + \delta) \setminus \{x_0\}. \tag{1.67}$$

Then

$$\lim_{x \to x_0} \frac{f(x)}{g(x)} = \lim_{x \to x_0} \frac{f'(x)}{g'(x)}, \tag{1.68}$$

provided the limit on right-hand side exists.

Proof. (i) Without loss of generality, we assume that $g'(x_0) > 0$. For $\varepsilon = \frac{g'(x_0)}{2} > 0$, there exists a $\delta > 0$ such that

$$\left| \frac{g(x) - g(x_0)}{x - x_0} - g'(x_0) \right| < \frac{g'(x_0)}{2}, \quad 0 < |x - x_0| < \delta,$$

which leads to (note $g(x_0) = 0$)

$$\frac{g(x)}{x - x_0} > \frac{g'(x_0)}{2}, \quad 0 < |x - x_0| < \delta.$$

Hence,

$$g(x) \neq 0, \quad 0 < |x - x_0| < \delta.$$

Then

$$\lim_{x \to x_0} \frac{f(x)}{g(x)} = \lim_{x \to x_0} \frac{f(x) - f(x_0)}{g(x) - g(x_0)} = \lim_{x \to x_0} \frac{\frac{f(x) - f(x_0)}{x - x_0}}{\frac{g(x) - g(x_0)}{x - x_0}} = \frac{f'(x_0)}{g'(x_0)}.$$

(ii) First of all, by (1.63), we claim that (1.61) holds for the given $\delta > 0$. In fact, if there exists an $\bar{x} \in (x_0, x_0 + \delta)$ such that $g(\bar{x}) = 0$, then applying Rolle's Theorem on $[x_0, \bar{x}]$, we have some $\bar{y} \in (x_0, \bar{x})$ such that $g'(\bar{y}) = 0$ which contradicts the assumption. Hence, our claim holds. Now, by Cauchy Mean Value Theorem, for any $x \in (x_0, x_0 + \delta)$, there exists $\xi \in (x_0, x)$ such that

$$\frac{f(x)}{g(x)} = \frac{f(x) - f(x_0)}{g(x) - g(x_0)} = \frac{f'(\xi)}{g'(\xi)}.$$

Hence,

$$\lim_{x \to x_0^+} \frac{f(x)}{g(x)} = \lim_{x \to x_0^+} \frac{f'(\xi)}{g'(\xi)} = \lim_{x \to x_0^+} \frac{f'(x)}{g'(x)},$$

provided the right-hand side exists.

The proofs of (iii) and (iv) are similar. □

The above result is very useful for finding the limit of the expression $\frac{f(x)}{g(x)}$ as $x \to x_0$ when $\lim_{x \to x_0} f(x) = \lim_{x \to x_0} g(x) = 0$.

Exercises

1. Let $\mathbb{R}^{n \times m}$ be the set of all $(n \times m)$ real matrices. Let $A \in \mathbb{R}^{n \times n}$ and $f(x) = x^\top A x$, for all $x \in \mathbb{R}^n$. Show that $f(\cdot)$ is Fréchet differentiable on $\mathbb{R}^n$. Find $f_x(x)$.

2. Let $f(x) = \|x\|_p \equiv \left(\sum\limits_{k=1}^{m} |x_k|^p \right)^{\frac{1}{p}}$, for any $x \in \mathbb{R}^m$, with $p \in [1, \infty)$. Then $f(\cdot)$ is Fréchet differentiable at any $x \in \mathbb{R}^m \backslash \{0\}$. Discuss the situation at $x = 0$.

3. Suppose it is known that $\lim\limits_{x \to 0} \dfrac{\sin x}{x} = 1$. Derive the following formulas by definition:

$$(\sin x)' = \cos x, \qquad (\cos x)' = -\sin x.$$

4. Let

$$f(x) = \begin{cases} |x|^\alpha \sin \dfrac{1}{|x|^\beta}, & x \neq 0, \\ 0, & x = 0. \end{cases}$$

Find conditions on $\alpha, \beta \in \mathbb{R}$ so that $f(\cdot)$ has derivative at $x = 0$.

5. Let $f : \mathbb{R} \to \mathbb{R}$ be differentiable. (i) If $f(x)$ is an even function, then $f'(x)$ is an odd function; (ii) If $f(x)$ is an odd function, then $f'(x)$ is an even function.

6. Let $f, g : \mathbb{R} \to \mathbb{R}$. Suppose $f(x)$ is differentiable at x_0 and $g(x)$ is not differentiable at x_0. What about the differentiability of $f(x) + g(x)$ and $f(x)g(x)$ at x_0? Why?

7. Let

$$f(x) = \begin{cases} \dfrac{1}{2^{2n}}, & x = \dfrac{1}{2^n}, \quad n \geqslant 1, \\ 0, & \text{otherwise.} \end{cases}$$

Prove or disprove the differentiability of $f(x)$ at $x = 0$. Is $f(x)$ continuous at $x = 0$? Why?

8. Let

$$f(x) = \begin{cases} x^2 \sin \dfrac{1}{x}, & x \neq 0, \\ 0, & x = 0. \end{cases}$$

Show that $f'(x)$ exists but not continuous.

9*. Let

$$f(x, y) = \begin{cases} \dfrac{x^p y^q}{|x|^n + |y|^m}, & (x, y) \neq (0, 0), \\ 0, & (x, y) = (0, 0), \end{cases}$$

where $m, n, p, q > 0$. Find conditions on m, n, p, q such that $f(x, y)$ is Gâteaux differentiable, has a gradient, and Fréchet differentiable, respectively.

10. Prove Theorem 1.15 parts (iii) and (iv).

11. Let $f, g : \mathbb{R} \to \mathbb{R}$ be continuous and differentiable with

$$\lim_{x \to \infty} f(x) = \lim_{x \to \infty} g(x) = 0.$$

Let

$$g'(x) \neq 0, \quad x \in \mathbb{R}, \qquad \lim_{x \to \infty} \frac{f'(x)}{g'(x)} = L,$$

for some $L \in \mathbb{R}$. Show that

$$\lim_{x \to \infty} \frac{f(x)}{g(x)} = L.$$

2 Fréchet Differentiability

In the previous section, we have seen that, in general, existence of a gradient (which implies the existence of all first order partial derivatives) does not implies the Fréchet differentiability. In this section, we would like to study the following problem: If $f(\cdot)$ admits all the first order partial derivatives at x_0, under what additional condition(s), will $f(\cdot)$ be Fréchet differentiable at x_0? We have the following result.

Theorem 2.1. *Let $f : G \to \mathbb{R}$ admit all the first order partial derivatives f_{x^i} $(1 \leqslant i \leqslant m)$ in a neighborhood of $x_0 \in G$, and all these partial derivatives are continuous at x_0. Then $f(\cdot)$ is Fréchet differentiable at x_0.*

Proof. Let us first look at the case $m = 2$. We write $x = (\xi, \eta)^\top$, and $x_0 = (\xi_0, \eta_0)^\top$. Since $\xi \mapsto f(\xi, \eta_0 + \eta)$ admits a derivative in a neighborhood of ξ_0 (for $|\eta|$ small enough), by Lagrange Mean Value Theorem, there exists a $\theta \in (0, 1)$ such that, for $|\xi|$ small,

$$f(\xi_0 + \xi, \eta_0 + \eta) - f(\xi_0, \eta_0 + \eta) = f_\xi(\xi_0 + \theta\xi, \eta_0 + \eta)\xi.$$

On the other hand, since $\eta \mapsto f(\xi_0, \eta)$ admits a derivative at η_0, one has

$$f(\xi_0, \eta_0 + \eta) - f(\xi_0, \eta_0) = f_\eta(\xi_0, \eta_0)\eta + R(\eta; \eta_0),$$

with

$$\lim_{|\eta| \to 0} \frac{|R(\eta; \eta_0)|}{|\eta|} = 0.$$

Consequently,

$$f(x_0 + x) - f(x_0) = f(\xi_0 + \xi, \eta_0 + \eta) - f(\xi_0, \eta_0)$$
$$= f_\xi(\xi_0 + \theta\xi, \eta_0 + \eta)\xi + f_\eta(\xi_0, \eta_0)\eta + R(\eta; \eta_0)$$
$$\equiv (f_\xi(x_0), f_\eta(x_0)) \cdot x + \widetilde{R}(x, x_0, \theta),$$

with

$$\widetilde{R}(x, x_0, \theta) = [f_\xi(\xi_0 + \theta\xi, \eta_0 + \eta) - f_\xi(\xi_0, \eta_0)]\xi + R(\eta; \eta_0).$$

Clearly, (note $x = (\xi, \eta)^\top$ and the continuity of $f_\xi(\cdot)$)

$$\lim_{\|x\| \to 0} \frac{\|\widetilde{R}(x, x_0, \theta)\|}{\|x\|} \leqslant \lim_{\|x\| \to 0} \frac{\|R(\eta; \eta_0)\|}{\|x\|}$$
$$+ \lim_{\|x\| \to 0} \frac{\|f_\xi(\xi_0 + \theta\xi, \eta_0 + \eta) - f_\xi(\xi_0, \eta_0)\| \, \|\xi\|}{\|x\|} = 0.$$

This proves the Fréchet differentiability of $f(\cdot)$ at $x_0 = (\xi_0, \eta_0)^\top$.

In the above, we have used the continuity of the partial derivative $f_\xi(\xi, \eta)$ near (ξ_0, η_0), and the Fréchet differentiability of $\eta \mapsto f(\xi_0, \eta)$ at η_0.

Next, we use induction. Suppose our conclusion holds for the case of $G \subseteq \mathbb{R}^{m-1}$. Then we write $x = (\xi, \eta)$ and $x_0 = (\xi_0, \eta_0)$ with $\xi, \xi_0 \in \mathbb{R}$ and $\eta, \eta_0 \in \mathbb{R}^{m-1}$. Now, $f_\xi(\xi, \eta)$ is continuous near (ξ_0, η_0), and by induction hypothesis, $\eta \mapsto f(\xi_0, \eta)$ is Fréchet differentiable. Hence, copying the above proof line-by-line, we can get the proof of our conclusion. □

A careful observation of the above proof suggests that we actually have a little more general result.

Theorem 2.2. *Let*

$$\begin{cases} \Gamma \equiv \{i_1, \cdots, i_k\} \subseteq \{1, 2, \cdots, m\}, \\ \Lambda \equiv \{j_1, \cdots, j_{m-k}\} = \{1, 2, \cdots, m\} \setminus \Gamma. \end{cases} \tag{2.1}$$

Write $x = (\xi, \eta)$ where $\xi \equiv (x^{i_1}, \cdots, x^{i_k})^\top$ and $\eta \equiv (x^{j_1}, \cdots, x^{j_{m-k}})^\top$. Let $f : G \to \mathbb{R}$ admit all the first order partial derivatives at $x_0 \equiv (\xi_0, \eta_0)$; For each $i \in \Gamma$, $f_{x^i}(x)$ exists in a neighborhood of x_0 and continuous; and the map $\eta \mapsto f(\xi_0, \eta)$ is Fréchet differentiable at η_0. Then $f(\cdot)$ is Fréchet differentiable at x_0.

Proof. Our condition says that $f_\xi(\xi, \eta)$ is continuous near (ξ_0, η_0), and $\eta \mapsto f(\xi_0, \eta)$ is Fréchet differentiable. Hence, the above proof can be applied. $\square$

We point out that in the above Theorem 2.2, if we only assume that $\xi \mapsto f(\xi, \eta_0)$ and $\eta \mapsto f(\xi_0, \eta)$ are Fréchet differentiable at ξ_0 and η_0, separately, then we might not have the Fréchet differentiability of $f(\cdot)$ at $x_0 \equiv (\xi_0, \eta_0)$. In fact, Examples 1.2 and 1.5 all gives

$$f_x(0,0) = f_y(0,0) = 0.$$

Thus, $x \mapsto f(x, 0)$ and $y \mapsto f(0, y)$ are all Fréchet differentiable at 0. But the corresponding functions $f(x, y)$ are not Fréchet differentiable at $(0, 0)$.

However, we should point out that the conditions imposed in the above theorems are only sufficient for the Fréchet differentiability. The following example illustrate this point.

Example 2.3. Let $m = 2$ and

$$f(x, y) = \begin{cases} (x^2 + y^2) \sin \dfrac{1}{x^2 + y^2}, & (x, y) \neq 0, \\ 0, & (x, y) = 0. \end{cases}$$

Then we have

$$f_x(x, y) = \begin{cases} 2x \sin \dfrac{1}{x^2 + y^2} - \dfrac{2x}{x^2 + y^2} \cos \dfrac{1}{x^2 + y^2}, & (x, y) \neq 0, \\ 0, & (x, y) = 0, \end{cases}$$

and

$$f_y(x, y) = \begin{cases} 2y \sin \dfrac{1}{x^2 + y^2} - \dfrac{2y}{x^2 + y^2} \cos \dfrac{1}{x^2 + y^2}, & (x, y) \neq 0, \\ 0, & (x, y) = 0. \end{cases}$$

Clearly, both f_x and f_y are discontinuous at $(0, 0)$. However,

$$\lim_{\|(x,y)\| \to 0} \frac{|f(x, y) - f(0, 0) - (0, 0) \cdot (x, y)|}{\|(x, y)\|}$$

$$= \lim_{x^2 + y^2 \to 0} \sqrt{x^2 + y^2} \left| \sin \frac{1}{x^2 + y^2} \right| = 0,$$

which means that $f(\cdot)$ is Fréchet differentiable at $(0, 0)$ with $\nabla f(0, 0) = (0, 0)$.

Despite of the above example, Theorem 2.1 is still one of the most commonly used sufficient conditions for the considered function to be Fréchet differentiable. We now introduce the following definition.

Definition 2.4. Let $G \subseteq \mathbb{R}^m$. Function $f : \bar{G} \to \mathbb{R}$ is said to be C^1 (on $\bar{G}$) if $f(\cdot)$ is continuous on $\bar{G}$, admits all the first order partial derivatives on G, and all of these partial derivatives are continuous on G and can be continuously extended to ∂G.

By Theorem 2.1, we see that if $f : \bar{G} \to \mathbb{R}$ is C^1, then it is Fréchet differentiable on G, and the Fréchet derivative $f_x(\cdot)$, which admits a representation (1.19), in terms of partial derivatives, is continuous. From now on, we will only consider Fréchet differentiability. Thus, for simplicity, we will omit the word "Fréchet" in Fréchet differentiability. Also, we will not distinguish $f_x(x)$ and $(\nabla f)(x)$ hereafter when the function is differentiable.

Exercises

1. Let $\| \cdot \| = \| \cdot \|_2$ be the ℓ^2-norm in $\mathbb{R}^m$. Define

$$f(x,y) = \begin{cases} \dfrac{|x^\top y|^p}{\|x\| + \|y\|}, & x \neq 0, \text{ or } y \neq 0, \\ 0, & x = y = 0, \end{cases}$$

with $x, y \in \mathbb{R}^m$. Discuss the differentiability of $f(x,y)$ at $(x,y) = (0,0)$ for different $p > 0$.

2. Let $f : \mathbb{R}^m \to \mathbb{R}$ be Lipschitz continuous, i.e., the following holds:

$$|f(x) - f(y)| \leqslant L\|x - y\|, \qquad \forall x, y \in \mathbb{R}^m,$$

for some $L > 0$. Suppose $f(x_0) = 0$. Then $x \mapsto f(x)^2$ is Fréchet differentiable at x_0.

3. Let $f : \mathbb{R}^m \to \mathbb{R}$ be convex, i.e.,

$$f(\lambda x + (1 - \lambda)y) \leqslant \lambda f(x) + (1 - \lambda)f(y), \qquad \forall x, y \in \mathbb{R}^m, \ \lambda \in (0,1).$$

Suppose $f(\cdot)$ is also differentiable. Then

$$f(x) + f_x(x)(y - x) \leqslant f(y), \qquad \forall x, y \in \mathbb{R}^m.$$

4. Let $f : \mathbb{R}^2 \to \mathbb{R}$ be defined by the following:

$$f(x,y) = \begin{cases} \dfrac{x \sin y}{y}, & y \neq 0, \\ x, & y = 0. \end{cases}$$

Prove or disprove that $f(x,y)$ is Fréchet differentiable at $(0,0)$.

3   Inverse Function and Implicit Function Theorems

Let $A \in \mathbb{R}^{m \times m}$ and $b \in \mathbb{R}^m$. Consider

$$f(x) = Ax + b, \qquad x \in \mathbb{R}^m.$$

If A^{-1} exists, then $f(\cdot)$ admits an *inverse function*

$$f^{-1}(y) = A^{-1}(y - b), \qquad y \in \mathbb{R}^m,$$

with

$$(f^{-1})_y(y) = A^{-1} = [f_x(f^{-1}(y))]^{-1}. \tag{3.1}$$

We now would like to extend the above to nonlinear case. The following is called the *Inverse Function Theorem*.

Theorem 3.1. *Let $f : G \to \mathbb{R}^m$ be C^1 and $x_0 \in G$ such that $\det[f_x(x_0)] \neq 0$. Then there exist open sets $U \subseteq G$ containing x_0 and $V \subseteq \mathbb{R}^m$ containing $f(x_0)$ such that $f(U) = V$, and there is an inverse function $f^{-1} : V \to U$ of $f(\cdot)$, which is C^1 in V with*

$$(f^{-1})_y(y) = [f_x(f^{-1}(y))]^{-1}, \qquad \forall y \in V. \tag{3.2}$$

Proof. Replacing $f(\cdot)$ by $[f_x(x_0)]^{-1}f(\cdot)$, if necessary, we may assume that $f_x(x_0) = I$ (the identity matrix of order m). Now, by (1.50), we have

$$\|x - \bar{x}\| - \|f(x) - f(\bar{x})\| \leqslant \|f(x) - x - [f(\bar{x}) - \bar{x}]\|$$
$$\leqslant \|f_x(\bar{x} + \theta(x - \bar{x})) - I\| \, \|x - \bar{x}\|,$$

for some $\theta \in (0, 1)$. Consequently,

$$\{1 - \|f_x(\bar{x} + \theta(x - \bar{x})) - I\|\} \, \|x - \bar{x}\| \leqslant \|f(x) - f(\bar{x})\|.$$

Since $f_x(\cdot)$ is continuous, and $f_x(x_0) = I$, we can find a $\delta > 0$ such that

$$\|f_x(x) - I\| < \frac{1}{2}, \qquad \forall x \in \bar{B}(x_0, \delta).$$

Hence, for some $\theta' \in (0, 1)$,

$$\frac{1}{2}\|x - \bar{x}\| \leqslant \|f(x) - f(\bar{x})\| \leqslant \|f_x(\bar{x} + \theta'(x - \bar{x}))\|\|x - \bar{x}\|$$
$$\leqslant \left(1 + \|f_x(\bar{x} + \theta'(x - \bar{x})) - I\|\right)\|x - \bar{x}\| \leqslant \frac{3}{2}\|x - \bar{x}\|, \tag{3.3}$$
$$\forall x, \bar{x} \in \bar{B}(x_0, \delta).$$

The above implies that $f(\cdot)$ is one-to-one on $\bar{B}(x_0, \delta)$. Further, since $\partial B(x_0, \delta)$ is compact, we have from (3.3) that

$$\sigma = \min_{x \in \partial B(x_0,\delta)} \|f(x) - f(x_0)\| \geqslant \frac{1}{2} \min_{x \in \partial B(x_0,\delta)} \|x - x_0\| = \frac{\delta}{2} > 0. \quad (3.4)$$

Let

$$V = \left\{ y \in \mathbb{R}^m \mid \|y - f(x_0)\| < \frac{\sigma}{2} \right\} \equiv B_{\frac{\sigma}{2}}(f(x_0)). \quad (3.5)$$

Now, for any fixed $y \in V$, we want to find a unique $\bar{x} \in B(x_0, \delta)$ such that $f(\bar{x}) = y$. To this end, we consider continuous function $h(x) = \|f(x) - y\|^2$ on the closed ball $\bar{B}(x_0, \delta)$. It attains its minimum at some $\bar{x} \in \bar{B}(x_0, \delta)$. We claim that $\bar{x} \notin \partial B(x_0, \delta)$. In fact, if $\bar{x} \in \partial B(x_0, \delta)$, then by (3.4)

$$\frac{\sigma}{2} > \|f(x_0) - y\| = \sqrt{h(x_0)} \geqslant \sqrt{h(\bar{x})}$$

$$= \|f(\bar{x}) - y\| \geqslant \|f(\bar{x}) - f(x_0)\| - \|y - f(x_0)\| > \sigma - \frac{\sigma}{2} = \frac{\sigma}{2},$$

which is a contradiction. This means that $\bar{x} \in B(x_0, \delta)$ (an open ball). Hence, by Fermat's theorem,

$$0 = h_x(\bar{x}) = 2f_x(\bar{x})(f(\bar{x}) - y).$$

Consequently, by the invertibility of $f_x(\bar{x})$ (since $\bar{x} \in B(x_0, \delta)$), we obtain that

$$f(\bar{x}) = y.$$

It follows from (3.3) that such an $\bar{x}$ must be uniquely determined by y. Hence, $y \mapsto \bar{x}$ defines the inverse $f^{-1}(\cdot)$ of $f(\cdot)$. If we let $U = f^{-1}(V)$, then by the continuity of $f(\cdot)$, U must be open, and $f(U) = V$, $f : U \to V$ is one-to-one and onto.

Next, we prove the differentiability of $f^{-1}(\cdot)$. To this end, we take any $\bar{x} \in U$. By the differentiability of $f(\cdot)$, we have

$$f(x) = f(\bar{x}) + f_x(\bar{x})(x - \bar{x}) + R(x, \bar{x}),$$

with

$$\lim_{\|x - \bar{x}\| \to 0} \frac{\|R(x, \bar{x})\|}{\|x - \bar{x}\|} = 0.$$

Denote $\bar{y} = f(\bar{x})$ and $y = f(x)$, with $x \neq \bar{x}$ (so that $y \neq \bar{y}$). Then the above becomes

$$y - \bar{y} - f_x(f^{-1}(\bar{y}))[f^{-1}(y) - f^{-1}(\bar{y})] = R(f^{-1}(y), f^{-1}(\bar{y})).$$

By applying $-[f_x(f^{-1}(\bar{y}))]^{-1}$ on the both sides of the above, one has

$$f^{-1}(y) - f^{-1}(\bar{y}) - [f_x(f^{-1}(\bar{y}))]^{-1}(y - \bar{y})$$
$$= -[f_x(f^{-1}(\bar{y}))]^{-1} R(f^{-1}(y), f^{-1}(\bar{y})) \equiv \widetilde{R}(y, \bar{y}).$$

Note that (taking into account (3.3))

$$\lim_{y \to \bar{y}} \frac{\|\widetilde{R}(y, \bar{y})\|}{\|y - \bar{y}\|} = \lim_{y \to \bar{y}} \frac{\|[f_x(f^{-1}(\bar{y}))]^{-1} R(f^{-1}(y), f^{-1}(\bar{y}))\|}{\|y - \bar{y}\|}$$
$$\leqslant \|[f_x(f^{-1}(\bar{y}))]^{-1}\| \left\{ \lim_{x \to \bar{x}} \frac{\|R(x, \bar{x})\|}{\|x - \bar{x}\|} \frac{\|x - \bar{x}\|}{\|f(x) - f(\bar{x})\|} \right\}$$
$$\leqslant 2\|[f_x(f^{-1}(\bar{y}))]^{-1}\| \left\{ \lim_{x \to \bar{x}} \frac{\|R(x, \bar{x})\|}{\|x - \bar{x}\|} \right\} = 0.$$

This means that $f^{-1}(\cdot)$ is differentiable and (2.1) holds, which further implies that $f^{-1}(\cdot)$ is C^1. ∎

The following is called the *Implicit Function Theorem*.

Theorem 3.2. Let $G \subseteq \mathbb{R}^k \times \mathbb{R}^\ell$, and $f : G \to \mathbb{R}^\ell$ be C^1 such that $f(x_0, y_0) = 0$ and $f_y(x_0, y_0) \in \mathbb{R}^{\ell \times \ell}$ is invertible. Then there exists a $\delta > 0$ and a unique map $g : B(x_0, \delta) \to \mathbb{R}^\ell$ such that

$$g(x_0) = y_0, \tag{3.6}$$

and

$$f(x, g(x)) = 0, \qquad \forall x \in B(x_0, \delta). \tag{3.7}$$

Moreover,

$$g_x(x) = -f_y(x, g(x))^{-1} f_x(x, g(x)), \qquad x \in B(x_0, \delta). \tag{3.8}$$

Proof. Let

$$F(x, y) = (x, f(x, y)), \qquad (x, y) \in G.$$

Then $F : G \to \mathbb{R}^{k+\ell}$ is differentiable, such that

$$F_{(x,y)}(x, y) = \begin{pmatrix} I & 0 \\ f_x(x, y) & f_y(x, y) \end{pmatrix}, \qquad (x, y) \in G.$$

Thus,

$$\det[F_{(x,y)}(x_0, y_0)] = \det[f_y(x_0, y_0)] \neq 0.$$

Also,

$$(u_0, v_0) \equiv F(x_0, y_0) = (x_0, f(x_0, y_0)) = (x_0, 0). \tag{3.9}$$

Hence, by Theorem 3.1, there exist open sets U containing (x_0, y_0) and V containing $(x_0, 0)$, such that $F : U \to V$ is one-to-one and onto, and $F^{-1} : V \to U$ is C^1. Now, if we denote

$$(x, y) = F^{-1}(u, v) = (\varphi(u, v), \psi(u, v)), \qquad (u, v) \in V, \tag{3.10}$$

then

$$(u, v) = F \circ F^{-1}(u, v) = F(\varphi(u, v), \psi(u, v))$$
$$= (\varphi(u, v), f(\varphi(u, v), \psi(u, v))), \qquad (u, v) \in V.$$

This means that

$$x = \varphi(u, v) = u, \qquad \forall (u, v) \in V, \tag{3.11}$$

which implies that

$$v = f(u, \psi(u, v)), \qquad \forall (u, v) \in V. \tag{3.12}$$

Now, by letting

$$g(x) = \psi(x, 0), \qquad \forall x \in B(x_0, \delta), \tag{3.13}$$

for some $\delta > 0$, we have that $g(\cdot)$ is C^1, and

$$f(x, g(x)) = f(x, \psi(x, 0)) = f(u, \psi(u, 0)) = 0, \tag{3.14}$$

which gives (3.7). Also, from (3.9)–(3.10), (3.11), and (3.13), we have

$$y_0 = \psi(u_0, v_0) = \psi(x_0, 0) = g(x_0),$$

which gives (3.6). Finally, by chain rule,

$$0 = f_x(x, g(x)) + f_y(x, g(x))g_x(x).$$

Hence, (3.8) follows. $\qquad\qquad\qquad\qquad\qquad\qquad\qquad\qquad\qquad\square$

Let us present an example.

Example 3.3. Suppose $G \subseteq \mathbb{R}^m$ and $f : G \to \mathbb{R}$. Write $x = (\xi, x^m) \in \mathbb{R}^m$. Let $x_0 = (\xi_0, x_0^m) \in G$ such that

$$f_{x^m}(x_0) \neq 0,$$

and

$$f(x_0) \equiv f(\xi_0, x_0^m) = c.$$

Then there exists a $\delta > 0$ and a unique $g : B^{m-1}(\xi_0, \delta) \subseteq \mathbb{R}^{m-1} \to \mathbb{R}$ such that

$$g(\xi_0) = x_0^m, \quad f(\xi, g(\xi)) = 0, \qquad \forall \xi \in B^{m-1}(\xi_0, \delta).$$

The function $x^m = g(\xi)$ is called the *level set surface* whose graph coincides with the following:

$$f^{-1}(c) \equiv (f = c) \equiv \{x \in G \mid f(x) = c\}.$$

Next, let $G \subseteq \mathbb{R}^m$ be a domain, and $f : G \to \mathbb{R}^m$ be C^1. By Proposition 1.14, we know that

$$f_x(x) = 0, \qquad \forall x \in G, \tag{3.15}$$

if and only if $f(\cdot)$ is a constant vector. Now, we would like to ask what will happen if instead of (3.15), we have the following condition:

$$\det[f_x(x)] \equiv \frac{\partial f(x)}{\partial x} \equiv \frac{\partial(f^1, \cdots, f^m)}{\partial(x^1, \cdots, x^m)}(x) = 0, \quad \forall x \in G. \tag{3.16}$$

The above implies that for each $x \in G$, there exists a non-zero vector $\xi(x) \in \mathbb{R}^m$ such that

$$f_x(x)^\top \xi(x) = 0. \tag{3.17}$$

It is natural to imagine that the components of $f(\cdot)$ should have some special relations. To get some feeling, let us look at the following situation. Suppose there exists a C^1 function $\theta : \mathbb{R}^m \to \mathbb{R}$ such that

$$\theta(f(x)) = 0, \qquad x \in G, \tag{3.18}$$

and

$$\theta_y(y) \neq 0, \qquad y \in f(G),$$

then, by the chain rule, we have

$$f_x(x)^\top \theta_y(f(x))^\top = 0, \qquad x \in G,$$

which is of the form (3.17) with $\xi(x) = \theta_y(f(x))^\top$. In particular, if

$$\theta(y) = \lambda^\top y, \qquad \forall y \in \mathbb{R}^m, \qquad \text{for some } \lambda \in \mathbb{R}^m \setminus \{0\},$$

then (3.18) means that $f^1, \cdots, f^m$ are linearly dependent, and (3.17) holds with $\xi(x) = \lambda$. Inspired by this, when (3.18) holds for some C^1 map $\theta(\cdot)$, we say that the components $f^1(\cdot), \cdots, f^m(\cdot)$ are *functionally dependent* on G. Hence, the above means that when $f^1(\cdot), \cdots, f^m(\cdot)$ are functionally dependent, then (3.16) holds. The following gives a kind of converse of such a statement.

Theorem 3.4. *Let $G \subseteq \mathbb{R}^m$ be a bounded domain and let $f : G \to \mathbb{R}^m$ be C^1 on G. Suppose (3.16) holds in a neighborhood $B(x_0, \delta) \subseteq G$ of some point $x_0 \in G$. Moreover,*

$$\text{rank}\,[f_x(x_0)] = m - 1,$$

meaning that if we write the $(m \times m)$ matrix $f_x(x) = (a_1, \cdots, a_m)$, then the set $\{a_1, \cdots, a_m\}$ is linearly dependent, and there exists at least one subset of it containing $(m - 1)$ vectors that is linearly independent. Then there exists a neighborhood of x_0 and a C^1 function $\theta : \mathbb{R}^m \to \mathbb{R}$ such that (3.18) holds near x_0.

Proof. Without loss of generality, we assume that

$$\frac{\partial(f^1, \cdots, f^{m-1})}{\partial(x^1, \cdots, x^{m-1})}(x) \neq 0, \qquad x \in B(x_0, \delta).$$

Denote

$$\xi = (x^1, \cdots, x^{m-1})^\top, \quad \xi_0 = (x_0^1, \cdots, x_0^{m-1})^\top,$$
$$\varphi(x) \equiv \varphi(\xi, x^m) = (f^1(x), \cdots, f^{m-1}(x))^\top.$$

Let $F : \mathbb{R}^{m-1} \times \mathbb{R} \times \mathbb{R}^{m-1} \to \mathbb{R}^{m-1}$ be defined by

$$F(\xi, x^m, u) = \varphi(\xi, x^m) - u, \quad \text{with} \quad u_0 = \varphi(x_0) = \varphi(\xi_0, x_0^m).$$

Clearly,

$$F(\xi_0, x_0^m, u_0) = 0.$$

Consider the equation $F(\xi, x^m, u) = 0$ (for ξ). Since

$$F_\xi(\xi_0, x_0^m, u_0) = \varphi_\xi(\xi_0, x_0^m) = \varphi_\xi(x_0)$$

is invertible, by the Implicit Function Theorem, we have a locally defined C^1 function $g : \mathcal{O} \to \mathbb{R}^{m-1}$ with $\mathcal{O} \subseteq \mathbb{R}^m$ being a neighborhood of $(x_0^m, \varphi(x_0))$, such that

$$F(g(x^m, u), x^m, u) \equiv \varphi(g(x^m, u), x^m) - u = 0, \qquad \forall (x^m, u) \in \mathcal{O}. \quad (3.19)$$

On the other hand, for any fixed x^m near x_0^m, the map $\xi \mapsto \varphi(\xi, x^m)$ satisfies the condition of the inverse function theorem near ξ_0. Therefore, we see that the above obtained function $u \mapsto g(x^m, u)$ is the inverse of $\xi \mapsto \varphi(\xi, x^m)$. Namely,

$$g(x^m, \varphi(\xi, x^m)) = \xi, \qquad \forall (\xi, x^m) \text{ near } (\xi_0, x_0^m) = x_0. \quad (3.20)$$

Now, differentiating (3.19) with respect to x^m, we get (suppressing (x^m, u))

$$0 = \varphi_\xi g_{x^m} + \varphi_{x^m}, \qquad \forall (x^m, u) \in \mathcal{O},$$

which leads to

$$g_{x^m} = -(\varphi_\xi)^{-1} \varphi_{x^m}, \qquad \forall (x^m, u) \in \mathcal{O},$$

Next, we differentiate the map $x^m \mapsto f^m(g(x^m, u), x^m)$ to get:

$$[f^m(g(x^m, u), x^m)]_{x^m} = f_\xi^m g_{x^m} + f_{x^m}^m$$
$$= f_{x^m}^m - f_\xi^m(\varphi_\xi)^{-1}\varphi_{x^m} = \det[f_x] = 0.$$

This means that $h(u) \equiv f^m(g(x^m, u), x^m)$ is independent of x^m. Then by defining

$$\theta(f) = h(f^1, \cdots, f^{m-1}) - f^m,$$

and taking into account (3.20), we obtain that in a neighborhood of x_0,

$$\theta(f(x)) = h(\varphi(x)) - f^m(x) = f^m(g(x^m, \varphi(x)), x^m) - f^m(x)$$
$$= f^m(\xi, x^m) - f^m(x) = 0,$$

proving our conclusion. □

We now would like to look at some geometric feature of the gradient for scalar-valued functions. To this end, we first introduce more notions.

Definition 3.5. Let $f : G \to \mathbb{R}$ be given.

(i) Let $x_0 \in f^{-1}(c)$ and $u \in \mathbb{R}^m$ be a unit vector. Then the straight line $x = x_0 + tu$ ($t \in \mathbb{R}$) is a *tangent line* of the *level set* $f^{-1}(c)$ at x_0 if

$$\lim_{h \to 0} \frac{\rho(x_0 + hu, f^{-1}(c))}{h} = 0, \tag{3.21}$$

where

$$\rho(x, f^{-1}(c)) \triangleq \inf_{y \in f^{-1}(c)} \|y - x\|, \tag{3.22}$$

is the distance from x to $f^{-1}(c)$. In this case, u is called a *tangent vector* of $f^{-1}(c)$ at x_0.

(ii) Let $x_0 \in f^{-1}(c)$. A plane, denoted by $\mathcal{T}_{x_0}$ is called the *tangent plane* of $f^{-1}(c)$ at x_0 if any straight line on this plane passing through x_0 is a tangent line of $f^{-1}(c)$ at x_0.

According to the above definition, if the level set $f^{-1}(c)$ is given by a circular cone with the vertex x_0, then $f^{-1}(c)$ does not have tangent lines at x_0, since (3.21) cannot hold for any direction. Hence, it does not have a tangent plane either. Note that the above definition can be extended to a more general case by replacing $f^{-1}(c)$ by any surface S.

The following result is concerned with tangent vectors.

Proposition 3.6. *Let $f : G \to \mathbb{R}$ and $g : [0, T] \to G$ be C^1 such that*

$$f(g(t)) = c, \qquad t \in [0, T]. \tag{3.23}$$

Then $\dot{g}(t) \equiv \frac{d}{dt}g(t)$ is a tangent vector of $f^{-1}(c)$ at $g(t) \in f^{-1}(c)$.

Proof. Let $t_0 \in [0, T]$ be fixed and $x_0 = g(t_0) \in f^{-1}(c)$. We know that $x = x_0 + t\dot{g}(t_0)$ ($t \in \mathbb{R}$) is the tangent line of the curve $g(\cdot)$ at $g(t_0) \equiv x_0$. Hence, one has

$$\lim_{h \to 0} \frac{\|g(t_0 + h) - x_0 - h\dot{g}(t_0)\|}{h} = 0.$$

By (3.22), we obtain

$$\varlimsup_{h\to 0} \frac{\rho(x_0 + h\dot{g}(t_0), f^{-1}(c))}{h} \leqslant \varlimsup_{h\to 0} \frac{\|g(t_0 + h) - x_0 - h\dot{g}(t_0)\|}{h} = 0.$$

This proves our claim. □

Next result gives the existence of tangent planes, for which we need to use the implicit function theorem.

Theorem 3.7. *Let* $f : G \to \mathbb{R}$ *be* C^1, *and let* $x_0 \in f^{-1}(c)$ *such that* $\nabla f(x_0) \neq 0$. *Then the plane passing through* x_0 *with normal vector* $\nabla f(x_0)$ *is the tangent plane of* $f^{-1}(c)$ *at* x_0. *The equation for this tangent plane is given by*

$$\nabla f(x_0)(x - x_0) = 0. \tag{3.24}$$

Proof. It suffices to show that any vector $u \perp \nabla f(x_0)^\top$ must be a tangent vector of $f^{-1}(c)$ at x_0. We now prove this. Consider the following equation (for $v \in \mathbb{R}$):

$$F(t, v) \equiv f(x_0 + tu + v\nabla f(x_0)^\top) - c = 0.$$

Clearly,

$$F(0, 0) = f(x_0) - c = 0,$$

and

$$\frac{\partial}{\partial v} F(t, v)\Big|_{(t,v)=(0,0)} = \nabla f(x_0)\nabla f(x_0)^\top = \|\nabla f(x_0)\|^2 > 0.$$

Hence, by implicit function theorem (Theorem 3.2), one can find a function $v(\cdot)$ defined on $[-\delta, \delta]$ for some $\delta > 0$, such that $v(0) = 0$ and

$$F(t, v(t)) \equiv f(x_0 + tu + v(t)\nabla f(x_0)^\top) = 0, \qquad t \in [-\delta, \delta].$$

Differentiating the above with respect to t, and evaluating at $t = 0$, we have (noting $u \perp \nabla f(x_0)^\top$)

$$0 = \nabla f(x_0)u + \|\nabla f(x_0)\|^2\dot{v}(0) = \|\nabla f(x_0)\|^2\dot{v}(0).$$

Thus, it is necessary that $\dot{v}(0) = 0$. Then with $g(t) = x_0 + tu + v(t)\nabla f(x_0)^\top$, one has

$$\begin{cases} f(g(t)) = c, & t \in [-\delta, \delta], \\ g(0) = x_0, & \dot{g}(0) = u. \end{cases} \tag{3.25}$$

Hence, by Proposition 3.6, we see that u is a tangent vector of $f^{-1}(c)$ at x_0. This proves our theorem. □

The above theorem is very useful in finding tangent planes for the level surface $f^{-1}(c)$ of given function $f(\cdot)$.

To conclude this section, we introduce the following notions.

Definition 3.8. Let $g : G \to \mathbb{R}$ be C^1 and $S \subseteq \mathbb{R}^m$ be a surface. Let $x_0 \in S$ and $\mathbf{N}(x_0)$ be a unit normal vector of S at x_0, by definition, $\mathbf{N}(x_0)$ is a normal vector of the tangent plane of S at x_0. The directional derivative of $g(\cdot)$ at point x_0 in the normal direction $\mathbf{N}(x_0)$ is called the *normal derivative* of $g(\cdot)$ on the surface S (with respect to $\mathbf{N}(x_0)$), denoted by $\frac{\partial g}{\partial n}$.

According to the above definition, it is clear that

$$\frac{\partial g}{\partial n}(x_0) = \nabla g(x_0)\mathbf{N}(x_0). \tag{3.26}$$

Further, if the surface S is closed and given by $f^{-1}(c)$ for some $f : G \to \mathbb{R}$ and $c \in \mathbb{R}$ (for example, a sphere), then the unit normal vector $\mathbf{N}(x_0)$ is usually taken to be the one pointing outside of the enclosed domain, which is called the *outward unit normal vector*. Thus, if the "inside" domain is characterized by $(f < c)$, then $\frac{\nabla f(x_0)^\top}{\|\nabla f(x_0)\|}$ is the outward unit normal vector, and

$$\frac{\partial g}{\partial n}(x_0) = \nabla g(x_0)\frac{\nabla f(x_0)^\top}{\|\nabla f(x_0)\|}. \tag{3.27}$$

Exercises

1. Let $f : \mathbb{R}^m \to \mathbb{R}^m$ be C^1 such that for each $x \in \mathbb{R}^m$, $f_x(x) \in \mathbb{R}^{m \times m}$ is an invertible matrix. Then for any open set $G \subseteq \mathbb{R}^m$, $f(G)$ is open.

2. Let $f, g, h : \mathbb{R} \to \mathbb{R}$ be continuously differentiable such that $f(\cdot)$ is non-decreasing, $g(x) > 0$ for all $x \in [0, \infty)$. Moreover,

$$f(1) + g(1) + h(1) = 0.$$

Then there exists a $\delta > 0$ such that for some continuously differentiable function $\varphi : (1 - \delta, 1 + \delta) \to \mathbb{R}$, the following holds:

$$x f(\varphi(x)) + g(x)\varphi(x) + h(x) = 0, \qquad x \in (1 - \delta, 1 + \delta).$$

3. Let $\varphi : \mathbb{R}^m \to \mathbb{R}$ be a C^1 convex function such that $\varphi_x(x) \neq 0$ for all $x \in \mathbb{R}^m$. Let

$$G \equiv \left\{ x \in \mathbb{R}^m \mid \varphi(x) \leqslant 0 \right\} \neq \varnothing.$$

Let $\bar{x} \in \partial G$. Then there exists a $k = 1, 2, \cdots, m$, a $\delta > 0$, and a C^1 function $\theta : \Gamma \subseteq \mathbb{R}^{m-1} \to \mathbb{R}$, with

$$\Gamma = \{(x_1, \cdots, x_{k-1}, x_{k+1}, \cdots, x_m) \mid$$
$$(x_1, \cdots, x_{k-1}, x_k, x_{k+1}, \cdots, x_m) \in B_\delta(0) + \bar{x}\}$$

such that

$$\varphi(x_1, \cdots, x_{k-1}, \theta(x_1, \cdots, x_{k-1}, x_{k+1}, \cdots, x_m), x_{k+1}, \cdots, x_m) = 0,$$

$$\forall (x_1, \cdots, x_{k-1}, x_{k+1} \cdots, x_m) \in \Gamma.$$

Give a geometric interpretation of the above result.

4 Higher Order Derivatives

For a differentiable function $f : G \to \mathbb{R}$, $f_x : G \to \mathbb{R}^m$ is a vector-valued function. We may further consider the differentiability of $f_x(\cdot)$. This leads to the second order derivatives for the original function $f(\cdot)$. Since $f_x(\cdot)$ is not just a vector-valued function, it is the gradient of a function, therefore, we expect to have some additional feature of the (partial) derivative/gradient of $f_x(\cdot)$. A natural question is: For $i \neq j$, is it true that

$$f_{x^i x^j}(x) = f_{x^j x^i}(x)? \tag{4.1}$$

Unfortunately, the above might fail in general. Here is an example.

Example 4.1. Let $m = 2$ and

$$f(x, y) = \begin{cases} xy \left(\dfrac{x^2 - y^2}{x^2 + y^2} \right), & (x, y) \neq 0, \\ 0, & (x, y) = 0. \end{cases} \tag{4.2}$$

A straightforward calculation shows that:

$$f_x(0, y) = -y, \qquad f_y(x, 0) = x. \tag{4.3}$$

Consequently,

$$f_{xy}(0, 0) = -1 \neq 1 = f_{yx}(0, 0). \tag{4.4}$$

Thus, for this function, (4.1) fails.

The following gives some sufficient conditions under which (4.1) holds.

Theorem 4.2. (Clairaut) *Let $f : G \to \mathbb{R}$ such that all first order partial derivatives $f_{x^i}(\cdot)$ exist in a neighborhood of $x_0 \in G$. Moreover, all of these partial derivatives are differentiable at x_0. Then (4.1) holds.*

Proof. It suffices to prove the case $m = 2$. Let $(a, b) \in G$ be the point that we are interested in. Define

$$\begin{cases} \varphi(x) = f(x, b + h) - f(x, b), \\ \psi(y) = f(a + h, y) - f(a, y). \end{cases} \tag{4.5}$$

Clearly, $\varphi(\cdot)$ and $\psi(\cdot)$ are differentiable. Hence, by Lagrange Mean Value Theorem (see Theorem 1.13, special case (ii)), there exist $\theta_1, \theta_2 \in (0,1)$ such that

$$\varphi(a+h) - \varphi(a) = \varphi'(a+\theta_1 h)h = [f_x(a+\theta_1 h, b+h) - f_x(a+\theta_1 h, b)]h, \quad (4.6)$$

and

$$\psi(b+h) - \psi(b) = \psi'(b+\theta_2 h)h = [f_y(a+h, b+\theta_2 h) - f_y(a+h, b)]h. \quad (4.7)$$

Now, by the differentiability of f_x at (a,b), we further have

$$\varphi(a+h) - \varphi(a) = \left\{ \left[f_x(a,b) + f_{xx}(a,b)\theta_1 h + f_{xy}(a,b)h + \varepsilon_1 |h| \right] \right.$$
$$\left. - \left[f_x(a,b) + f_{xx}(a,b)\theta_1 h + \varepsilon_1' |h| \right] \right\} h \qquad (4.8)$$
$$= f_{xy}(a,b)h^2 + (\varepsilon_1 + \varepsilon_1')|h|h,$$

where $\varepsilon_1, \varepsilon_1' \to 0$ as $|h| \to 0$. Likewise, by the differentiability of f_y at (a,b), we have

$$\psi(b+h) - \psi(b) = f_{yx}(a,b) + (\varepsilon_2 + \varepsilon_2')|h|h, \qquad (4.9)$$

with $\varepsilon_2, \varepsilon_2' \to 0$ as $|h| \to 0$. On the other hand,

$$\varphi(a+h) - \varphi(a) = f(a+h, b+h) - f(a+h, b) - f(a, b+h) + f(a, b)$$
$$= \psi(b+h) - \psi(b). \qquad (4.10)$$

Consequently,

$$f_{xy}(a,b) - f_{yx}(a,b) = (\varepsilon_2 + \varepsilon_2' - \varepsilon_1 - \varepsilon_1')\frac{|h|}{h} \to 0. \qquad (4.11)$$

Hence, our conclusion follows. $\qquad \square$

Combining the above with Theorem 4.1, we obtain the following:

Corollary 4.3. *Suppose $f : G \to \mathbb{R}$ admits all the first and second order partial derivatives in a neighborhood of $x_0 \in G$ and all of them are continuous at x_0. Then (4.1) holds at $x = x_0$.*

Definition 4.4. *Let $G \subseteq \mathbb{R}^m$. Function $f : \bar{G} \to \mathbb{R}$ is said to be C^2 on $\bar{G}$ if $f(\cdot)$ is continuous on $\bar{G}$, admits all the first and second order partial derivatives and all of these derivatives are continuous on G and can be continuously extended to ∂G.*

If $f(\cdot)$ is C^2, we denote

$$f_{xx}(x) = \begin{pmatrix} f_{x^1x^1}(x) & f_{x^1x^2}(x) & \cdots & f_{x^1x^m}(x) \\ f_{x^2x^1}(x) & f_{x^2x^2}(x) & \cdots & f_{x^2x^m}(x) \\ \vdots & \vdots & \cdots & \vdots \\ f_{x^mx^1}(x) & f_{x^mx^2}(x) & \cdots & f_{x^mx^m}(x) \end{pmatrix}, \tag{4.12}$$

which is an $(m \times m)$ symmetric matrix. We call the above matrix the *Hessian* of function $f(\cdot)$.

We may further define higher order Fréchet derivatives and higher order partial derivatives.

Definition 4.5. Let $G \subseteq \mathbb{R}^m$. Function $f : \bar{G} \to \mathbb{R}$ is said to be C^k on $\bar{G}$ if $f(\cdot)$ is continuous on $\bar{G}$, admits all the partial derivatives up to order k, and all of these partial derivatives are bounded and continuous on G, and can be continuously extended to ∂G. Further, if $f(\cdot)$ is C^k on $\bar{G}$ for any k, we say that $f(\cdot)$ is C^∞ on $\bar{G}$.

Theorem 4.6. (Taylor) *Let $f : [a,b] \to \mathbb{R}$ be C^{k+1} and $x_0 \in (a,b)$. Then for any $x \in [a,b]$, there exists a ξ between x_0 and x such that*

$$f(x) = f(x_0) + \frac{f'(x_0)}{1!}(x-x_0) + \frac{f''(x_0)}{2!}(x-x_0)^2 + \cdots$$
$$+ \frac{f^{(k)}(x_0)}{k!}(x-x_0)^k + \frac{f^{(k+1)}(\xi)}{(k+1)!}(x-x_0)^{k+1}. \tag{4.13}$$

Proof. Fix an $x \in (a,b)$. Let

$$\begin{cases} \bar{f}(t) = f(t) + \dfrac{f'(t)}{1!}(x-t) + \dfrac{f''(t)}{2!}(x-t)^2 + \cdots + \dfrac{f^{(k)}(t)}{k!}(x-t)^k, \\ \bar{g}(t) = (x-t)^{k+1}. \end{cases}$$

Then by Cauchy Mean-Value Theorem, we have some ξ between x_0 and x such that

$$\frac{\bar{f}'(\xi)}{\bar{g}'(\xi)} = \frac{\bar{f}(x) - \bar{f}(x_0)}{\bar{g}(x) - \bar{g}(x_0)}.$$

Note that

$$\bar{f}'(t) = f'(t) - f'(t) + \frac{f''(t)}{1!}(x-t) - \frac{f''(t)}{1!}(x-t) + \cdots$$
$$+ \frac{f^{(k)}(t)}{(k-1)!}(x-t)^{k-1} - \frac{f^{(k)}(t)}{(k-1)!}(x-t)^{k-1} + \frac{f^{(k+1)}(t)}{k!}(x-t)^k$$
$$= \frac{f^{(k+1)}(t)}{k!}(x-t)^k,$$

and

$$\bar{g}'(t) = -(k+1)(x-t)^k.$$

Thus,

$$-\frac{f^{(k+1)}(\xi)}{(k+1)!} = \frac{\bar{f}'(\xi)}{\bar{g}'(\xi)} = \frac{\bar{f}(x) - \bar{f}(x_0)}{\bar{g}(x) - \bar{g}(x_0)} = \frac{f(x) - \bar{f}(x_0)}{-\bar{g}(x_0)}$$

$$= \frac{1}{-(x-x_0)^{k+1}}\Big[f(x) - \Big(f(x_0) + \frac{f'(x_0)}{1!}(x-x_0)$$

$$+ \frac{f''(x_0)}{2!}(x-x_0)^2 + \cdots + \frac{f^{k)}(x_0)}{k!}(x-x_0)^k\Big)\Big].$$

Hence, we obtain (4.6). □

Now, we look at the multi-variable case.

Corollary 4.7. *Let $f : G \to \mathbb{R}$ be C^{k+1} in a neighborhood of $x_0 \in G$. Then*

$$f(x) = \sum_{i=0}^k \sum_{|\alpha|=i} \frac{1}{\alpha!} D^\alpha f(x_0)(x-x_0)^\alpha$$

$$+ \sum_{|\alpha|=k+1} \frac{1}{\alpha!} D^\alpha f(x_0 + \theta(x-x_0))(x-x_0)^\alpha,$$

(4.14)

for some $\theta \in (0,1)$, where $\alpha = (\alpha_1, \alpha_2, \cdots, \alpha_m) \in (\mathbb{N} \cup \{0\})^m$ is called a multi-index, with $|\alpha| = \alpha_1 + \cdots + \alpha_m$, $\alpha! = \alpha_1!\alpha_2!\cdots\alpha_m!$ and

$$D^\alpha \varphi(x) = \partial_{x^1}^{\alpha_1} \cdots \partial_{x^m}^{\alpha_m} \varphi(x), \qquad \forall \alpha = (\alpha_1, \cdots, \alpha_m).$$

In particular,

$$f(x) = f(x_0) + f_x(x_0) \cdot (x-x_0)$$

$$+ \frac{1}{2}(x-x_0)^\top f_{xx}(x_0 + \theta(x-x_0))(x-x_0).$$

(4.15)

Proof. Fix $x_0 \in G$ and for any x near x_0, define

$$h(t) = f(x_0 + t(x - x_0)), \qquad t \in [-\delta, 1+\delta],$$

for some $\delta > 0$. Then $h(\cdot) \in C^{k+1}$. Applying the above Taylor Theorem, we have

$$h(1) - h(0) = h'(0) + \frac{h''(0)}{2!} + \cdots + \frac{h^{(k)}(0)}{k!} + \frac{h^{(k+1)}(\theta)}{(k+1)!},$$

for some $\theta \in (0,1)$. The above leads to (4.14) by some careful calculation. □

The above result leads to a refinement of the Fermat's theorem.

Proposition 4.8. Let $f : G \to \mathbb{R}$ be C^2 in a neighborhood of $x_0 \in G$.

(i) Suppose that $f(\cdot)$ attains a local maximum (resp. local minimum) at x_0. Then

$$f_x(x_0) = 0, \qquad (4.16)$$

and

$$\xi^\top f_{xx}(x_0)\xi \leqslant 0, \quad (\text{resp.} \geqslant 0), \quad \forall \xi \in \mathbb{R}^m. \qquad (4.17)$$

(ii) Suppose that (4.16) holds and

$$\xi^\top f_{xx}(x_0)\xi < 0, \qquad (\text{resp.} > 0), \quad \forall \xi \in \mathbb{R}^m \setminus \{0\}. \qquad (4.18)$$

Then x_0 is a local maximum (resp. minimum) of $f(\cdot)$.

Proof. (i) Suppose $f(\cdot)$ attains a local maximum at x_0. Then by Fermat's theorem and (4.15), we have

$$f(x) = f(x_0) + \frac{1}{2}(x - x_0)^\top f_{xx}(x_0 + \theta(x - x_0))(x - x_0) \leqslant f(x_0),$$

for all x near x_0, which implies that

$$\xi^\top f_{xx}(x_0 + \theta(x - x_0))\xi \leqslant 0, \qquad \forall \xi \in \mathbb{R}^m.$$

Then (4.17) follows from the continuity of $f_{xx}(\cdot)$ (by sending $x \to x_0$).

Similarly, we can prove the case that x_0 is a local minimum of $f(\cdot)$.

(ii) From (4.15), we see that when (4.16) and (4.18) hold, we have (making use of the continuity of $f_{xx}(\cdot)$)

$$f(x) \leqslant f(x_0), \qquad (4.19)$$

for all x near x_0. This means that x_0 is a local maximum of $f(\cdot)$. The conclusion for a local minimum can be proved similarly. $\qquad \square$

Exercises

1. Construct a smooth function $\varphi : \mathbb{R}^m \to \mathbb{R}$ such that

$$\varphi(x) = \begin{cases} 1, & \|x\|_2 \leqslant 1, \\ 0, & \|x\|_2 \geqslant 2. \end{cases}$$

2. Let $f : \mathbb{R}^m \to \mathbb{R}$ be C^2. Then $f(\cdot)$ is convex, i.e.,

$$f(\lambda x + (1 - \lambda)y) \leqslant \lambda f(x) + (1 - \lambda)f(y), \qquad \forall x, y \in \mathbb{R}^m, \ \lambda \in (0, 1),$$

if and only if the Hessian $f_{xx}(x)$ is positive semi-definite, denoted by $f_{xx}(x) \geqslant 0$, for any $x \in \mathbb{R}^m$.

3. (**Darboux Theorem**) Let $f : [a, b] \to \mathbb{R}$ have derivatives on $[a, b]$. Then for every y between $f'(a)$ and $f'(b)$, there exists a $c \in [a, b]$ such that $f'(c) = y$. (Hint: Consider $g(t) = f(t) - yt$.)

4. Let $f : \mathbb{R} \to \mathbb{R}$ satisfy the following: For some $\alpha > 1$,

$$|f(x) - f(y)| \leqslant K|x - y|^\alpha, \qquad \forall x, y \in \mathbb{R}.$$

Then $f(x) \equiv f(0)$ for all $x \in \mathbb{R}$.

5. Let $f : \mathbb{R} \to \mathbb{R}$ be differentiable and $\lim\limits_{x \to \infty} f'(x) = 0$. Then for any $a \in \mathbb{R}$ fixed, $\lim\limits_{x \to \infty} \left[f(x + a) - f(x)\right] = 0$.

6. Let $a, b \in \mathbb{R}$.

(i) Show that if n is even, equation $x^n + ax + b = 0$ has at most two distinct real roots;

(ii) Show that if n is odd, equation $x^n + ax + b = 0$ has at most three distinct real roots.

7. Suppose $a_0, a_1, \cdots, a_n \in \mathbb{R}$ such that

$$a_0 + \frac{a_1}{2} + \cdots + \frac{a_{n-1}}{n} + \frac{a_n}{n+1} = 0.$$

Then the following equation

$$a_n x^n + a_{n-1} x^{n-1} + \cdots + a_1 x + a_0 = 0$$

has a real root in $(0, 1)$.

8. Let $F(x) = f(x)g(x)$ with $f(x)$ and $g(x)$ being n-th time differentiable. Find a formula of $F^{(n)}(x)$ in terms of $f(x)$ and $g(x)$ and their derivatives. (Hint: Using induction.)

9. Let $f(\cdot)$ be C^2 on (a, b) and let $f''(x) \not\equiv 0$. Then for any $x \in (a, b)$ and $h > 0$ with $x + 2h \in (a, b)$,

$$f'(x) = \frac{f(x + 2h) - f(x)}{2h} - hf''(\xi),$$

for some $\xi \in (x, x + 2h)$. Further,

$$\sup_{x \in [a,b]} |f'(x)|^2 \leqslant 4 \Big(\sup_{x \in [a,b]} |f(x)| \Big) \Big(\sup_{x \in [a,b]} |f''(x)| \Big).$$

10. Let

$$f(x) = \begin{cases} e^{-\frac{1}{x^2}}, & x \neq 0, \\ 0, & x = 0. \end{cases}$$

Find the Taylor expansion of $f(\cdot)$ at $x = 0$. What conclusion can you draw from it?

Chapter 4

Riemann Integrals

1 Definition of Integrals

We recall $\mathbb{Z}$, the set of all integers, and let $\mathcal{A}^m$ be the set of all arrays α:

$$\alpha \equiv \left\{ (\alpha_1^{j_1}, \alpha_2^{j_2}, \cdots, \alpha_m^{j_m}) \mid \alpha_i^{j_i} \in \mathbb{R}, \ j_i \in \mathbb{Z}, \ 1 \leqslant i \leqslant m \right\} \subseteq \mathbb{R}^m, \qquad (1.1)$$

satisfying the following conditions:

$$\begin{cases} \alpha_i^j < \alpha_i^{j+1}, \qquad j \in \mathbb{Z}, \ 1 \leqslant i \leqslant m, \\ \lim_{j \to -\infty} \alpha_i^j = -\infty, \quad \lim_{j \to \infty} \alpha_i^j = \infty, \quad 1 \leqslant i \leqslant m, \\ \|\alpha\| \equiv \max_{1 \leqslant i \leqslant m} \sup_{j \in \mathbb{Z}} [\alpha_i^{j+1} - \alpha_i^j] < \infty. \end{cases} \qquad (1.2)$$

We call $\|\alpha\|$ the *mesh size* of α. For any $\alpha \in \mathcal{A}^m$, define

$$\mathcal{Q}^\alpha \triangleq \left\{ Q = \prod_{i=1}^m [\alpha_i^{j_i}, \alpha_i^{j_i+1}] \ \middle| \ (\alpha_1^{j_1}, \alpha_2^{j_2}, \cdots, \alpha_m^{j_m}) \in \alpha \right\}. \qquad (1.3)$$

Clearly,

$$\begin{cases} \mathbb{R}^m = \bigcup_{Q \in \mathcal{Q}^\alpha} Q, \\ Q^\circ \bigcap \tilde{Q} = \varnothing, \qquad \forall Q, \tilde{Q} \in \mathcal{Q}^\alpha, \ Q \neq \tilde{Q}. \end{cases} \qquad (1.4)$$

We call $\mathcal{Q}^\alpha$ the *rectangular partition* of $\mathbb{R}^m$ associated with the array α. For $Q = \prod_{i=1}^m [\alpha_i^{j_i}, \alpha_i^{j_i+1}] \in \mathcal{Q}^\alpha$, its *diameter*, denoted by diam (Q), is defined by the following:

$$\text{diam}\,(Q) = \left\{ \sum_{i=1}^m (\alpha_i^{j_i+1} - \alpha_i^{j_i})^2 \right\}^{\frac{1}{2}} \leqslant \sqrt{m}\,\|\alpha\|, \qquad (1.5)$$

and its *volume* (in $\mathbb{R}^m$), denoted by $|Q|$, is defined by

$$|Q| \equiv \prod_{i=1}^{m} (\alpha_i^{j_i+1} - \alpha_i^{j_i}) \leqslant \|\alpha\|^m. \tag{1.6}$$

If $\alpha, \beta \in \mathcal{A}^m$ such that $\alpha \subseteq \beta$, we call β a *refinement* of α. It is clear that

$$\|\beta\| \leqslant \|\alpha\|, \qquad \forall \alpha, \beta \in \mathcal{A}^m, \ \alpha \subseteq \beta. \tag{1.7}$$

Also, if $\alpha, \beta \in \mathcal{A}^m$, then $\alpha \cup \beta$ is a refinement of both α and β. Thus,

$$\|\alpha \cup \beta\| \leqslant \min\{\|\alpha\|, \|\beta\|\}, \qquad \forall \alpha, \beta \in \mathcal{A}^m. \tag{1.8}$$

Because of this, we see that for any sequence $\{\alpha_n\} \subseteq \mathcal{A}^m$, by defining

$$\beta_n = \bigcup_{k=1}^{n} \alpha_k, \qquad n \geqslant 1, \tag{1.9}$$

we obtain a sequence $\{\beta_n\} \subseteq \mathcal{A}^m$ such that

$$\alpha_n \subseteq \beta_n, \quad \beta_n \subseteq \beta_{n+1}, \qquad n \geqslant 1. \tag{1.10}$$

In what follows, whenever saying $\|\alpha\| \to 0$, we always mean that there exists a sequence $\{\alpha_n\} \subseteq \mathcal{A}^m$ such that $\alpha_n \subseteq \alpha_{n+1}$ for all $n \geqslant 1$ and $\|\alpha_n\| \to 0$.

Now, for a bounded domain $G \subseteq \mathbb{R}^m$, by our construction, for any $\alpha \in \mathcal{A}^m$, $\bar{G}$ and ∂G will be covered by finitely many subsets of $\mathcal{Q}^\alpha$, respectively. We will denote $\mathcal{Q}^\alpha(G)$ and $\mathcal{Q}^\alpha(\partial G)$ to be the smallest subsets of $\mathcal{Q}^\alpha$ covering $\bar{G}$ and ∂G, respectively. Also, we will denote

$$\mathcal{Q}^{\alpha,\circ}(G) = \mathcal{Q}^\alpha(G) \setminus \mathcal{Q}^\alpha(\partial G). \tag{1.11}$$

Let

$$\mathcal{Q}^\alpha(G) = \{Q_\ell^\alpha \mid 1 \leqslant \ell \leqslant \bar{\ell}_\alpha\}, \quad \mathcal{Q}^{\alpha,\circ}(G) = \{Q_\ell^\alpha \mid 1 \leqslant \ell \leqslant \ell_\alpha\}, \tag{1.12}$$

where $\ell_\alpha \leqslant \bar{\ell}_\alpha$. Thus,

$$\mathcal{Q}^\alpha(\partial G) = \{Q_\ell^\alpha \mid \ell_\alpha + 1 \leqslant \ell \leqslant \bar{\ell}_\alpha\}. \tag{1.13}$$

Note that $\mathcal{Q}^{\alpha,\circ}(G)$ covers most part (but not necessarily all) of G°. If we let

$$G_\ell^\alpha = \bar{G} \bigcap Q_\ell^\alpha, \qquad 1 \leqslant \ell \leqslant \bar{\ell}_\alpha, \ \alpha \in \mathcal{A}^m, \tag{1.14}$$

then

$$G^\alpha \equiv \bigcup_{Q \in \mathcal{Q}^{\alpha,\circ}(G)} Q \subseteq G^\circ \subseteq \bar{G} \subseteq \bigcup_{Q \in \mathcal{Q}^\alpha(G)} Q \equiv \tilde{G}^\alpha, \tag{1.15}$$

and

$$|\widetilde{G}^\alpha \setminus G^\alpha| = \sum_{Q \in \mathcal{Q}^\alpha(\partial G)} |Q|. \tag{1.16}$$

The following assumption will be important later.

(H1) Let $G \subseteq \mathbb{R}^m$ be a bounded domain such that

$$|\partial G| \equiv \lim_{\|\alpha\| \to 0} \sum_{Q \in \mathcal{Q}^\alpha(\partial G)} |Q| = 0. \tag{1.17}$$

Condition (1.17) is referred to as the *zero volume condition* of the boundary ∂G of G in $\mathbb{R}^m$. Now, we present a simple result concerning bounded domains, whose proof is pretty obvious.

Proposition 1.1. *Let $G \subseteq \mathbb{R}^m$ be a bounded domain. If we denote*

$$V_+^\alpha(G) = \sum_{Q \in \mathcal{Q}^\alpha(G)} |Q|, \qquad V_-^\alpha(G) = \sum_{Q \in \mathcal{Q}^{\alpha,\circ}(G)} |Q|, \tag{1.18}$$

then

$$V_+^\alpha(G) \geqslant V_-^\alpha(G), \qquad \forall \alpha \in \mathcal{A}^m, \tag{1.19}$$

and

$$V_+^\alpha(G) \geqslant V_+^\beta(G) \geqslant V_-^\beta(G) \geqslant V_-^\alpha(G), \qquad \forall \alpha, \beta \in \mathcal{A}^m, \ \alpha \subseteq \beta. \tag{1.20}$$

Consequently,

$$\lim_{\|\alpha\| \to 0} V_+^\alpha(G) = V_+(G) \geqslant V_-(G) = \lim_{\|\alpha\| \to 0} V_-^\alpha(G). \tag{1.21}$$

Further, if (1.17) holds, then

$$V_+(G) = V_-(G) \equiv |G|, \tag{1.22}$$

which is called the volume of G.

Note that if condition (1.17) were not assumed for a bounded domain $G \subseteq \mathbb{R}^m$, we would not be able to define the volume $|G|$ as (1.22).

Now, let $G \subseteq \mathbb{R}^m$ be a given bounded domain satisfying (H1), and let $f : \bar{G} \to \mathbb{R}$ be a bounded function (not necessarily continuous). For any $\alpha \in \mathcal{A}^m$, recalling (1.11)–(1.15), we let

$$\begin{cases} \Lambda_\ell^\alpha(f) = \sup_{x \in G_\ell^\alpha} f(x), \\ \lambda_\ell^\alpha(f) = \inf_{x \in G_\ell^\alpha} f(x), \end{cases} \qquad 1 \leqslant \ell \leqslant \bar{\ell}_\alpha, \quad \alpha \in \mathcal{A}^m. \tag{1.23}$$

Define *simple functions*

$$f_+^\alpha(x) = \sum_{\ell=1}^{\ell_\alpha} \Lambda_\ell^\alpha(f) \mathbf{1}_{Q_\ell^\alpha}(x), \quad f_-^\alpha(x) = \sum_{\ell=1}^{\ell_\alpha} \lambda_\ell^\alpha(f) \mathbf{1}_{Q_\ell^\alpha}(x), \quad x \in G^\alpha,$$

$$\bar{f}_+^\alpha(x) = \sum_{\ell=1}^{\bar{\ell}_\alpha} \Lambda_\ell^\alpha(f) \mathbf{1}_{Q_\ell^\alpha}(x), \quad \bar{f}_-^\alpha(x) = \sum_{\ell=1}^{\bar{\ell}_\alpha} \lambda_\ell^\alpha(f) \mathbf{1}_{Q_\ell^\alpha}(x), \quad x \in \widetilde{G}^\alpha. \tag{1.24}$$

Note that if Q_ℓ^α and $Q_{\ell'}^\alpha$ are next to each other, then

$$\Gamma_{\ell,\ell'}^\alpha \equiv Q_\ell^\alpha \cap Q_{\ell'}^\alpha \neq \varnothing.$$

Therefore, rigorously speaking, according to the above notation, for any $x \in \Gamma_{\ell,\ell'}^\alpha$,

$$f_+^\alpha(x) = \Lambda_\ell^\alpha(f) + \Lambda_{\ell'}^\alpha(f),$$

and it might be even more complicated if the point x is in the intersection of more than two rectangular sets (since $\mathbb{R}^m$ could have the dimension $m > 2$). But, instead, we define

$$f_+^\alpha(x) = \Lambda_\ell^\alpha(f) \vee \Lambda_{\ell'}^\alpha(f),$$

and misuse the notation in (1.24) for simplicity (and convenience). Likewise, for any $x \in \Gamma_{\ell,\ell'}^\alpha$, we define

$$f_-^\alpha(x) = \lambda_\ell^\alpha(f) \wedge \lambda_{\ell'}^\alpha(f).$$

The case of x being in the intersection of more than two rectangular sets can be treated similarly.

Now, we have

$$\begin{aligned} f_-^\alpha(x) \leqslant f(x) \leqslant f_+^\alpha(x), \quad & x \in G^\alpha \subseteq G^\circ, \\ \bar{f}_-^\alpha(x) \leqslant f(x) \leqslant \bar{f}_+^\alpha(x), \quad & x \in \bar{G} \subseteq \widetilde{G}^\alpha. \end{aligned} \tag{1.25}$$

Further, let us define

$$S_+^\alpha(f) = \sum_{\ell=1}^{\ell_\alpha} \Lambda_\ell^\alpha(f)|Q_\ell^\alpha|, \qquad S_-^\alpha(f) = \sum_{\ell=1}^{\ell_\alpha} \lambda_\ell^\alpha(f)|Q_\ell^\alpha|,$$

$$\bar{S}_+^\alpha(f) = \sum_{\ell=1}^{\bar{\ell}_\alpha} \Lambda_\ell^\alpha(f)|Q_\ell^\alpha|, \qquad \bar{S}_-^\alpha(f) = \sum_{\ell=1}^{\bar{\ell}_\alpha} \lambda_\ell^\alpha(f)|Q_\ell^\alpha|. \tag{1.26}$$

We call $S_+^\alpha(f)$ and $\bar{S}_+^\alpha(f)$ the *upper Riemann sums*, and $S_-^\alpha(f)$ and $\bar{S}_-^\alpha(f)$ the *lower Riemann sums* of $f(\cdot)$ on G^α and $\bar{G}$, respectively. We have the following result whose proof is straightforward.

Proposition 1.2. *Let $G \subseteq \mathbb{R}^m$ be a bounded domain and let $f : \bar{G} \to \mathbb{R}$ be a bounded function. Then*

$$S_+^\alpha(f) \geqslant S_-^\alpha(f), \quad \bar{S}_+^\alpha(f) \geqslant \bar{S}_-^\alpha(f), \qquad \forall \alpha \in \mathcal{A}^m, \tag{1.27}$$

and

$$\begin{cases} S_+^\alpha(f) \geqslant S_+^\beta(f) \geqslant S_-^\beta(f) \geqslant S_-^\alpha(f), \\ \bar{S}_+^\alpha(f) \geqslant \bar{S}_+^\beta(f) \geqslant \bar{S}_-^\beta(f) \geqslant \bar{S}_-^\alpha(f), \end{cases} \quad \forall \alpha, \beta \in \mathcal{A}^m, \ \alpha \subseteq \beta. \tag{1.28}$$

Consequently,

$$\lim_{\|\alpha\| \to 0} S_+^\alpha(f) = S_+(f) \geqslant S_-(f) = \lim_{\|\alpha\| \to 0} S_-^\alpha(f),$$
$$\lim_{\|\alpha\| \to 0} \bar{S}_+^\alpha(f) = \bar{S}_+(f) \geqslant \bar{S}_-(f) = \lim_{\|\alpha\| \to 0} \bar{S}_-^\alpha(f). \tag{1.29}$$

Note that if $M > 0$ is a bound of $f(\cdot)$, then

$$|\bar{S}_+^\alpha(f) - S_+^\alpha(f)| = \Big| \sum_{\ell = \ell_\alpha + 1}^{\bar{\ell}_\alpha} \Lambda_\ell^\alpha(f) |Q_\ell^\alpha| \Big| \leqslant M \sum_{\ell = \ell_\alpha + 1}^{\bar{\ell}_\alpha} |Q_\ell^\alpha|,$$

and

$$|\bar{S}_-^\alpha(f) - S_-^\alpha(f)| = \Big| \sum_{\ell = \ell_\alpha + 1}^{\bar{\ell}_\alpha} \lambda_\ell^\alpha(f) |Q_\ell^\alpha| \Big| \leqslant M \sum_{\ell = \ell_\alpha + 1}^{\bar{\ell}_\alpha} |Q_\ell^\alpha|.$$

Hence,

$$|\bar{S}_+(f) - S_+(f)| = \varlimsup_{\|\alpha\| \to 0} |\bar{S}_+^\alpha(f) - S_+^\alpha(f)| \leqslant M |\partial G|,$$
$$|\bar{S}_-(f) - S_-(f)| = \varlimsup_{\|\alpha\| \to 0} |\bar{S}_-^\alpha(f) - S_-^\alpha(f)| \leqslant M |\partial G|. \tag{1.30}$$

Consequently, under (H1), one has

$$\bar{S}_+(f) = S_+(f) \geqslant \bar{S}_-(f) = S_-(f).$$

We refer to $S_+(f)$ and $S_-(f)$ as the *upper* and *lower Riemann integrals* of $f(\cdot)$, respectively. The above result shows that if $G \subseteq \mathbb{R}^m$ is a domain satisfying (H1), then any bounded function on G admits upper and lower Riemann integrals. Then the following definition becomes very natural.

Definition 1.3. If

$$S_+(f) = \bar{S}_+(f) = S_-(f) = \bar{S}_-(f), \tag{1.31}$$

then we say that $f(\cdot)$ is *(Riemann) integrable* on G, and denote

$$S_+(f) = S_-(f) = \int_G f(x)dx, \tag{1.32}$$

call it the *definite (Riemann) integral* of $f(\cdot)$ on G. Function $f(\cdot)$ is called an *integrand*. The set of all bounded Riemann integrable functions on $\bar{G}$ is denoted by $L_R^\infty(G)$. When $G = (a, b)$ for some $a, b \in \mathbb{R}$ with $a < b$, we usually denote

$$\int_G f(x)dx = \int_a^b f(x)dx, \tag{1.33}$$

and call a and b the *lower* and *upper limits* of the definite Riemann integral.

Note that if we define

$$S_*^\alpha(f) = \sum_{\ell=1}^{\ell_\alpha} f(x_\ell^*)|Q_\ell^\alpha|, \tag{1.34}$$

with $x_\ell^* \in G_\ell^\alpha$, then

$$S_-^\alpha(f) \leqslant S_*^\alpha(f) \leqslant S_+^\alpha(f). \tag{1.35}$$

Thus, in the case $f(\cdot)$ is Riemann integrable, one has

$$\lim_{\|\alpha\| \to 0} S_*^\alpha(f) = \int_G f(x)dx. \tag{1.36}$$

The following result shows that the set of Riemann integrable functions is very large.

Theorem 1.4. *Let $G \subseteq \mathbb{R}^m$ satisfy* (H1). *Then any continuous function $f : \bar{G} \to \mathbb{R}$ is Riemann integrable on G.*

Proof. Since $f(\cdot)$ is continuous on the compact set $\bar{G}$, it is uniformly continuous. Thus, for any $\varepsilon > 0$, there exists a $\delta > 0$ such that

$$0 \leqslant \Lambda_\ell^\alpha(f) - \lambda_\ell^\alpha(f) < \varepsilon, \qquad \forall 1 \leqslant \ell \leqslant \bar{\ell}_\alpha, \quad \alpha \in \mathcal{A}^m, \ \|\alpha\| < \delta.$$

Next, we let

$$M = \max_{x \in \bar{G}} |f(x)|. \tag{1.37}$$

Then the following holds:

$$0 \leqslant \bar{S}_+^\alpha(f) - \bar{S}_-^\alpha(f) = \sum_{\ell=1}^{\bar{\ell}_\alpha} \left[\Lambda_\ell^\alpha(f) - \lambda_\ell^\alpha(f) \right] |Q_\ell^\alpha| \leqslant \varepsilon \sum_{\ell=1}^{\bar{\ell}_\alpha} |Q_\ell^\alpha|.$$

Hence, it follows that

$$0 \leqslant \bar{S}_+(f) - \bar{S}_-(f) = \lim_{\|\alpha\| \to 0} \bar{S}_+^\alpha(f) - \lim_{\|\alpha\| \to 0} \bar{S}_-^\alpha(f) \leqslant C\varepsilon, \tag{1.38}$$

for some constant $C > 0$. Similarly,

$$0 \leqslant S_+(f) - S_-(f) = \lim_{\|\alpha\| \to 0} S_+^\alpha(f) - \lim_{\|\alpha\| \to 0} S_-^\alpha(f) \leqslant C\varepsilon. \tag{1.39}$$

Combining the above with (1.30), we have the Riemann integrability of $f(\cdot)$ since $\varepsilon > 0$ is arbitrary. $\qquad\square$

Example 1.5. Let $G \subseteq \mathbb{R}^m$ satisfy (H1). Then the function $f(x) \equiv 1$ is Riemann integrable on G and

$$|G| = \int_G dx.$$

This is the volume of G.

We point out that if $f(\cdot)$ is not continuous on $\bar{G}$, the above theorem might not be true. Here is an example.

Example 1.6. Let $G = (0, 1)$ and

$$f(x) = \begin{cases} 1, & x \in \mathbb{Q} \cap [0, 1], \\ 0, & x \in [0, 1] \setminus \mathbb{Q}. \end{cases} \tag{1.40}$$

Recall that this is the *Dirichlet function*. For this function, we see that

$$\Lambda_\ell^\alpha(f) = 1, \quad \lambda_\ell^\alpha(f) = 0, \quad \forall 1 \leqslant \ell \leqslant \bar{\ell}_\alpha, \quad \alpha \in \mathcal{A}^m.$$

Hence,

$$S_+^\alpha(f) = 1, \quad S_-^\alpha(f) = 0, \quad \forall \alpha \in \mathcal{A}^m.$$

Consequently, (1.31) fails. This example also shows that a bounded function is not necessarily in $L_R^\infty(G)$.

Despite of the above example, the following result tells us that some discontinuous functions are still Riemann integrable.

Theorem 1.7. *Let $G \subseteq \mathbb{R}^m$ satisfy (H1). Let $f : \bar{G} \to \mathbb{R}$ be bounded and denote*

$$D(f) = \{x \in G \mid f(\cdot) \text{ is discontinuous at } x\}. \tag{1.41}$$

Suppose that

$$|D(f)| \equiv \lim_{\|\alpha\| \to 0} \sum_{Q \in \mathcal{Q}^\alpha(\overline{D(f)})} |Q| = 0. \tag{1.42}$$

Then $f(\cdot)$ is Riemann integrable.

Condition (1.42) means that the set at which $f(\cdot)$ is discontinuous has a zero volume in $\mathbb{R}^m$. Note that Dirichlet function (see Example 1.6) does not satisfy condition (1.42).

The proof of Theorem 1.7 is essentially the same as that of Theorem 1.5 in which we replace ∂G by $\partial G \cup D(f)$. The details are left to the readers.

The following corollary is very useful.

Corollary 1.8. *Let $G \subseteq \mathbb{R}^m$ be a bounded domain satisfying* (H1) *and $f(\cdot)$ be Riemann integrable on G. Then for any rectangular domain $Q \supseteq G$, the function $\widetilde{f}(\cdot)$ defined by*

$$\widetilde{f}(x) = \begin{cases} f(x), & x \in \bar{G}, \\ 0, & x \in Q \setminus \bar{G}, \end{cases} \tag{1.43}$$

is Riemann integrable on Q, and

$$\int_Q \widetilde{f}(x)dx = \int_G f(x)dx. \tag{1.44}$$

Proof. Using the same technique used in the above proof, treating ∂G as a part of $D(\widetilde{f})$, the discontinuous set of $\widetilde{f}(\cdot)$, we are able to obtain the conclusion. $\qquad\square$

The above result shows that when discussing Riemann integrals, one can assume that the bounded domain G is a rectangular, as long as (H1) holds.

Next result gives a nice approximation of bounded Riemann integrable functions, which will be useful later.

Proposition 1.9. *Let $G \subseteq \mathbb{R}^m$ be a bounded domain satisfying* (H1). *Let $f : G \to \mathbb{R}$ be bounded and Riemann integrable. Then for any $\varepsilon > 0$, there are continuous functions $g_{\pm}^{\varepsilon} : G \to R$ such that*

$$g_-^{\varepsilon}(x) \leqslant f(x) \leqslant g_+^{\varepsilon}(x), \qquad \forall x \in G, \tag{1.45}$$

and

$$\int_G [g_+^{\varepsilon}(x) - g_-^{\varepsilon}(x)]dx < \varepsilon. \tag{1.46}$$

Proof. First of all, for any $a' < a < b < b'$, we define

$$\psi_{a',a,b,b'}(s) = \frac{s - a'}{a - a'}\mathbf{1}_{[a',a)}(s) + \mathbf{1}_{[a,b]}(s) + \frac{b' - s}{b' - b}\mathbf{1}_{(b,b']}(s), \quad s \in \mathbb{R}.$$

Then $\psi_{a',a,b,b'} : \mathbb{R} \to \mathbb{R}$ is a continuous function. Now, for any rectangular $Q = \prod_{i=1}^m [a_i, b_i]$, we let

$$Q' = \prod_{i=1}^m [a_i', b_i'], \qquad a_i' < a_i < b_i < b_i', \quad 1 \leqslant i \leqslant m,$$

and define

$$\psi_{Q,Q'}(x) = \prod_{i=1}^m \psi_{a_i',a_i,b_i,b_i'}(x^i), \qquad x = (x^1, \cdots, x^m) \in \mathbb{R}^m.$$

Clearly, $\psi_{Q,Q'} \in C(\mathbb{R}^m, \mathbb{R})$ and

$$\psi_{Q,Q'}(x) = \begin{cases} 1, & x \in Q, \\ 0, & x \in \mathbb{R}^m \setminus Q'. \end{cases}$$

For given rectangular Q, let Q', Q'' be rectangulars satisfying

$$Q'' \subset Q^\circ \subset Q \subset (Q')^\circ.$$

We have

$$\mathbf{1}_{Q''}(x) \leqslant \psi_{Q'',Q}(x) \leqslant \mathbf{1}_Q(x) \leqslant \psi_{Q,Q'}(x) \leqslant \mathbf{1}_{Q'}(x), \qquad x \in \mathbb{R}^m,$$

which leads to

$$0 \leqslant \int_{Q'} \psi_{Q,Q'}(x)dx - |Q| \leqslant |Q' \setminus Q|,$$

$$0 \leqslant |Q| - \int_{Q} \psi_{Q'',Q}(x)dx \leqslant |Q \setminus Q''|.$$

Thus,

$$0 \leqslant \int_{Q'} [\psi_{Q,Q'}(x) - \psi_{Q'',Q}(x)]dx \leqslant |Q' \setminus Q| + |Q \setminus Q''|.$$

Now, let $f : G \to [0, M]$ be bounded non-negative and Riemann integrable. Then for any $\varepsilon > 0$, there exists an $\alpha \in \mathcal{A}^m$ such that

$$0 \leqslant f_-^\alpha(x) = \sum_{\ell=1}^{\ell_\alpha} \lambda_\ell^\alpha(f) \mathbf{1}_{Q_\ell^\alpha}(x) \leqslant f(x)$$

$$\leqslant \sum_{\ell=1}^{\ell_\alpha} \Lambda_\ell^\alpha(f) \mathbf{1}_{Q_\ell^\alpha}(x) = f_+^\alpha(x) \leqslant M, \qquad x \in G,$$

and

$$0 \leqslant S_+(f) - S_-(f) = \int_G [f_+^\alpha(x) - f_-^\alpha(x)]dx$$

$$= \sum_{\ell=1}^{\ell_\alpha} [\Lambda_+^\alpha(f) - \lambda_-^\alpha(f)]|Q_\ell^\alpha| < \frac{\varepsilon}{2}.$$

Let us take

$$(Q_\ell^\alpha)'' \subset (Q_\ell^\alpha)^\circ \subset Q_\ell^\alpha \subset [(Q_\ell^\alpha)']^\circ.$$

Define

$$g_+^\alpha(x) = \sum_{\ell=1}^{\ell_\alpha} \Lambda_\alpha(f)\psi_{Q_\ell^\alpha,(Q_\ell^\alpha)'}(x), \quad g_-^\alpha(x) = \sum_{\ell=1}^{\ell_\alpha} \lambda_\alpha(f)\psi_{(Q_\ell^\alpha)'',Q_\ell^\alpha}(x).$$

Then

$$0 \leqslant g_-^\alpha(x) \leqslant f_-^\alpha(x) \leqslant f(x) \leqslant f_+^\alpha(x) \leqslant g_+^\alpha(x) \leqslant M, \quad x \in G,$$

and

$$\int_{\mathbb{R}^m} [g_+^\alpha(x) - g_-^\alpha(x)]dx$$

$$= \sum_{\ell=1}^{\ell_\alpha} \left\{ \Lambda_\ell^\alpha(f) \int_{\mathbb{R}^m} \psi_{Q_\ell^\alpha,(Q_\ell^\alpha)'}(x)dx - \lambda_\ell^\alpha(f) \int_{\mathbb{R}^m} \psi_{(Q_\ell^\alpha)'',Q_\ell^\alpha}(x) \right\}dx$$

$$= \sum_{\ell=1}^{\ell_\alpha} \left\{ \Lambda_\ell^\alpha(f) \left[\int_{\mathbb{R}^m} \psi_{Q_\ell^\alpha,(Q_\ell^\alpha)'}(x)dx - |Q_\ell^\alpha| \right] \right.$$

$$\left. + \lambda_\ell^\alpha(f) \left[|Q_\ell^\alpha| - \int_{\mathbb{R}^m} \psi_{(Q_\ell^\alpha)'',Q_\ell^\alpha}(x)dx \right] + \sum_{\ell=1}^{\ell_\alpha} [\Lambda_\ell^\alpha(f) - \lambda_\ell^\alpha(f)|Q_\ell^\alpha|] \right\}$$

$$\leqslant \sum_{\ell=1}^{\ell_\alpha} \left\{ \Lambda_\ell^\alpha(f)|(Q_\ell^\alpha)' \setminus Q_\ell^\alpha| + \lambda_\ell^\alpha(f)|Q_\ell^\alpha \setminus (Q_\ell^\alpha)''| \right\} + S_+^\alpha(f) - S_-^\alpha(f)$$

$$\leqslant M \sum_{\ell=1}^{\ell_\alpha} \left\{ |(Q_\ell^\alpha)' \setminus Q_\ell^\alpha| + |Q_\ell^\alpha \setminus (Q_\ell^\alpha)''| \right\} + \frac{\varepsilon}{2}.$$

Then by taking $(Q_\ell^\alpha)', (Q_\ell^\alpha)''$ suitably, we obtain our conclusion

For general case, we have $f(\cdot) = f^+(\cdot) - f^-(\cdot)$ with both $f^+(\cdot)$ and $f^-(\cdot)$ being non-negative. From the above, we can find (non-negative) continuous functions $g_+^+(\cdot), g_-^+(\cdot), g_+^-(\cdot), g_-^-(\cdot)$ such that

$$g_-^+(x) \leqslant f^+(x) \leqslant g_+^+(x), \quad g_-^-(x) \leqslant f^-(x) \leqslant g_+^-(x), \quad x \in G,$$

and

$$\int_{\mathbb{R}^m} [g_+^+(x) - g_-^+(x)]dx + \int_{\mathbb{R}^m} [g_+^-(x) - g_-^-(x)]dx < \varepsilon.$$

Then we have

$$g_-(x) \equiv g_-^+(x) - g_+^-(x) \leqslant f^+(x) - f^-(x)$$

$$= f(x) \leqslant g_+^+(x) - g_-^-(x) \equiv g_+(x),$$

and

$$\int_{\mathbb{R}^m} [g_+(x) - g_-(x)] dx = \int_{\mathbb{R}^m} [g_+^+(x) - g_-^-(x) - g_-^+(x) + g_+^-(x)] dx < \varepsilon.$$

This completes the proof. ☐

Next, we define the Riemann integral of possibly unbounded functions on possibly unbounded domains.

Definition 1.10. Let $G \subseteq \mathbb{R}^m$ be an unbounded domain with the property that

$$\lim_{M \to \infty} |\partial G_M(x_0)| = 0, \qquad \forall x_0 \in \mathbb{R}^m, \tag{1.47}$$

where

$$G_M(x_0) = \{x \in G \mid |x - x_0| < M\} \equiv G \cap B(x_0, M).$$

Let $f : G \to \mathbb{R}$ be a possibly unbounded function. For $-N_1, N_2 > 0$, define

$$[f]_{N_1}^{N_2}(x) = \begin{cases} f(x), & N_1 < f(x) < N_2, \\ N_1, & f(x) \leqslant N_1, \qquad x \in G, \\ N_2, & f(x) \geqslant N_2, \end{cases}$$

which is a bounded function defined on G. Suppose $[f]_{N_1}^{N_2}(\cdot)$ is Riemann integrable on $G_M(x_0)$, for any $x_0 \in \mathbb{R}^m$, $M, -N_1, N_2 > 0$; and for any $\varepsilon > 0$, $x_0 \in \mathbb{R}^m$, there exist $M_0, N_0 > 0$ such that

$$\left| \int_{G_M(x_0)} [f]_{N_1}^{N_2}(x) dx - \int_{G_{M'}(x_0)} [f]_{N_1'}^{N_2'}(x) dx \right| < \varepsilon, \tag{1.48}$$

$$\forall M, M' \geqslant M_0, \quad -N_1, -N_1', N_2, N_2' \geqslant N_0.$$

Then the following limit exists which is called the *Riemann integral* of the function $f(\cdot)$ over G:

$$\lim_{M, |N_1|, N_2 \to \infty} \int_{G_M(x_0)} [f]_{N_1}^{N_2}(x) dx = \int_G f(x) dx. \tag{1.49}$$

In what follows, we denote $L_R(G)$ to be the set of all Riemann integrable functions $f : G \to \mathbb{R}$, in the above sense, which could be unbounded and not necessarily continuous. Note that we should allow $|N_1| \neq N_2$ and $x_0 \in \mathbb{R}^m$ to be arbitrary. This will exclude the following function from being Riemann integrable:

$$f(x) = \begin{cases} \dfrac{1}{x}, & x \neq 0, \\ 0, & x = 0. \end{cases}$$

In fact, for this function, we have

$$\int_{-M}^{M} [f]_{-N}^{N}(x)dx = 0.$$

According to the above definition, if $G_0 \subseteq G$ are two domains, both satisfy (1.47), then $f(\cdot) \in L_R(G)$ implies $f(\cdot)|_{G_0} \in L_R(G_0)$, where $f(\cdot)|_{G_0}$ is the restriction of $f(\cdot)$ on G_0. Note that it might not be the case if G_0 is just a (possibly disconnected) sub-open set of G (see Problem 7 at the end of this section).

To conclude this section, we introduce the following definition.

Definition 1.11. Let $G \subseteq \mathbb{R}^m$ be a domain satisfying (1.47). Let $f : G \to \mathbb{R}$. If both $f(\cdot)$ and $|f(\cdot)|^p$ are Riemann integrable on G, for some $p \geqslant 1$, we say that $f(\cdot)$ is L^p-*Riemann integrable*; in the case $p = 1$, we say that $f(\cdot)$ is *absolutely Riemann integrable*.

It is possible that a function $f : G \to \mathbb{R}$ is not Riemann integrable but $|f(\cdot)|^p$ is Riemann integrable. Here is a simple example.

$$f(x) = \begin{cases} 1, & x \in \mathbb{Q} \cap [0,1], \\ -1, & x \in [0,1] \setminus \mathbb{Q}. \end{cases}$$

Hence, the above definition, excludes the above function to be L^p-Riemann integrable. We denote $L_R^p(G)$ to be the set of all L^p-Riemann integrable functions on G. Then it is clear that

$$L_R^1(G) \subsetneq L_R(G), \tag{1.50}$$

which means that absolute Riemann integrability implies Riemann integrability but not vice versa.

Exercises

1. Let $G \subseteq \mathbb{R}^m$ be a bounded domain with $|\partial G| = 0$ and $f : \bar{G} \to (0, \infty)$ be a continuous function. Then $\frac{1}{f(\cdot)}$ is Riemann integrable on G.

2. Let $f : [a, b] \to [0, \infty)$ be continuous. Then

$$\int_a^b f(x)dx \geqslant 0.$$

3. Let $f, g : [a, b] \to \mathbb{R}$ be continuous. Then

$$\left| \int_a^b f(x)g(x)dx \right| \leqslant \left(\int_a^b |f(x)|^2 dx \right)^{\frac{1}{2}} \left(\int_a^b |g(x)|^2 dx \right)^{\frac{1}{2}}.$$

(Hint: Consider $\displaystyle\int_a^b \left[f(x) + \lambda g(x)\right]^2 dx \geqslant 0$.)

4. Show that when (1.48) holds, the limit (1.49) exists.

5. Let

$$f(x) = \begin{cases} \dfrac{1}{q}, & x = \dfrac{p}{q} \in \mathbb{Q} \cap [0,1], \ q \neq 0, \\ 0, & x \in [\mathbb{R} \setminus \mathbb{Q}] \cap [0,1]. \end{cases}$$

Then $f(\cdot)$ is Riemann integrable on $[0,1]$. What is the value of $\int_0^1 f(x)dx$?

6. Let $f : [a,b] \to \mathbb{R}$ be Riemann integrable. Let $a_1, a_2, \cdots, a_n \in [a,b]$ be finitely many different points, and $c_1, c_2, \cdots, c_n \in \mathbb{R}$ be arbitrarily given. Define

$$g(x) = \begin{cases} f(x), & x \neq a_i, \ i = 1, 2, \cdots, n, \\ c_i, & x = a_i, \ i = 1, 2, \cdots, n. \end{cases}$$

Prove or disprove that $g(\cdot)$ is Riemann integrable.

7.* Let

$$f(x) = \begin{cases} \dfrac{\sin x}{x}, & x \in (0, \infty), \\ 1, & x = 0. \end{cases}$$

Show that $f(\cdot) \in L_R([0,\infty)) \setminus L_R^1([0,\infty))$. Moreover, there exists an open set $G_0 \subseteq [0,\infty)$ such that $\int_{G_0} f(x)dx = \infty$.

2 Properties of Integrals

The following result collects some basic properties of Riemann integrable functions.

Proposition 2.1. *Let $G \subseteq \mathbb{R}^m$ be a bounded domain satisfying* (H1).

(i) *For any $f(\cdot), g(\cdot) \in L_R(G)$, and $c \in \mathbb{R}$,*

$$\int_G [cf(x) + g(x)]dx = c\int_G f(x)dx + \int_G g(x)dx. \tag{2.1}$$

Consequently, $L_R(G)$ is a linear space. The same conclusion also holds for $L_R^p(G)$.

(ii) *If $f(\cdot), g(\cdot) \in L_R(G)$ and*

$$f(x) \leqslant g(x), \qquad \forall x \in G, \tag{2.2}$$

then

$$\int_G f(x)dx \leqslant \int_G g(x)dx. \tag{2.3}$$

Thus, the Riemann integral preserves the order of $\mathbb{R}$. *Consequently, for any* $f(\cdot) \in L_R^1(G)$,

$$\left| \int_G f(x)dx \right| \leqslant \int_G |f(x)|dx. \tag{2.4}$$

(iii) *Let* $f(\cdot), g(\cdot) \in L_R^\infty(G)$. *Then* $f(\cdot)g(\cdot)$, $|f(\cdot)|$, $f(\cdot) \vee g(\cdot)$, *and* $f(\cdot) \wedge g(\cdot)$ *are in* $L_R^\infty(G)$, *where recall that* $a \vee b = \max\{a, b\}$ *and* $a \wedge b = \min\{a, b\}$.

(iv) *Let* $\bar{G} = \bar{G}_1 \bigcup \bar{G}_2$ *with both* G_1 *and* G_2 *satisfying (H1), and*

$$G_1 \bigcap G_2 = \varnothing. \tag{2.5}$$

Let $f(\cdot) \in L_R^1(G)$. *Then* $f(\cdot) \in L_R^1(G_1)$, $f(\cdot) \in L_R^1(G_2)$, *and*

$$\int_G f(x)dx = \int_{G_1} f(x)dx + \int_{G_2} f(x)dx. \tag{2.6}$$

Proof. We leave the proofs of (i) and (ii) to the readers.

(iii) For any $\alpha \in \mathcal{A}^m$, we define $f_\pm^\alpha(\cdot)$ and $g_\pm^\alpha(\cdot)$ as in (1.24). Then

$$f_-^\alpha(x) \leqslant f(x) \leqslant f_+^\alpha(x), \quad g_-^\alpha(x) \leqslant g(x) \leqslant g_+^\alpha(x), \quad x \in G^\alpha. \tag{2.7}$$

This yields

$$f_-^\alpha(x) \vee g_-^\alpha(x) \leqslant f(x) \vee g(x) \leqslant f_+^\alpha(x) \vee g_+^\alpha(x), \quad x \in G^\alpha. \tag{2.8}$$

Hence,

$$
\begin{aligned}
f_-^\alpha(x) \vee g_-^\alpha(x) &\leqslant \sum_{\ell=1}^{\ell_\alpha} \lambda_\ell^\alpha(f \vee g) \mathbf{1}_{G_\ell^\alpha}(x) \\
&\leqslant \sum_{\ell=1}^{\ell_\alpha} \Lambda_\ell^\alpha(f \vee g) \mathbf{1}_{G_\ell^\alpha}(x) \leqslant f_+^\alpha(x) \vee g_+^\alpha(x), \quad x \in G^\alpha.
\end{aligned}
\tag{2.9}
$$

This leads to

$$
\begin{aligned}
\int_{G^\alpha} f_-^\alpha(x) \vee g_-^\alpha(x)dx &\leqslant S_-^\alpha(f \vee g) \leqslant S_+^\alpha(f \vee g) \\
&\leqslant \int_{G^\alpha} f_+^\alpha(x) \vee g_+^\alpha(x)dx.
\end{aligned}
\tag{2.10}
$$

By the Riemann integrability of $f(\cdot)$ and $g(\cdot)$, for any $\varepsilon > 0$, there exists a $\beta \in \mathcal{A}^m$ such that

$$-\varepsilon + \int_G f(x)dx \leqslant S_-^\alpha(f) \leqslant S_+^\alpha(f) \leqslant \int_G f(x)dx + \varepsilon,$$
$$-\varepsilon + \int_G g(x)dx \leqslant S_-^\alpha(g) \leqslant S_+^\alpha(g) \leqslant \int_G g(x)dx + \varepsilon, \qquad (2.11)$$
$$\forall \alpha \in \mathcal{A}^m, \ \alpha \supseteq \beta.$$

Let

$$0 \leqslant h^\alpha(\cdot) = f_+^\alpha(\cdot) - f_-^\alpha(\cdot) + g_+^\alpha(\cdot) - g_-^\alpha(\cdot). \qquad (2.12)$$

Then by (ii),

$$0 \leqslant \int_{G^\alpha} h^\alpha(x)dx$$
$$= \int_{G^\alpha} f_+^\alpha(x)dx - \int_{G^\alpha} f_-^\alpha(x)dx + \int_{G^\alpha} g_+^\alpha(x)dx - \int_{G^\alpha} g_-^\alpha(x)dx \qquad (2.13)$$
$$= S_+^\alpha(f) - S_-^\alpha(f) + S_+^\alpha(g) - S_-^\alpha(g) \leqslant 4\varepsilon.$$

Observe the following:

$$f_+^\alpha(x) = f_-^\alpha(x) + \left[f_+^\alpha(x) - f_-^\alpha(x)\right] \leqslant f_-^\alpha(x) + h^\alpha(x),$$

and

$$g_+^\alpha(x) = g_-^\alpha(x) + \left[g_+^\alpha(x) - g_-^\alpha(x)\right] \leqslant g_-^\alpha(x) + h^\alpha(x).$$

Therefore,

$$f_+^\alpha(x) \vee g_+^\alpha(x) \leqslant f_-^\alpha(x) \vee g_-^\alpha(x) + h^\alpha(x).$$

Hence,

$$0 \leqslant S_+^\alpha(f \vee g) - S_-^\alpha(f \vee g)$$
$$\leqslant \int_{G^\alpha} \left[f_+^\alpha(x) \vee g_+^\alpha(x) - f_-^\alpha(x) \vee g_-^\alpha(x)\right]dx \leqslant \int_{G^\alpha} h^\alpha(x)dx \leqslant 4\varepsilon.$$

Since $\varepsilon > 0$ is arbitrary, we obtain the Riemann integrability of $f \vee g$.

The Riemann integrability of $f \wedge g$ follows from that of $-[(-f) \vee (-g)]$.

Then we further obtain the Riemann integrability of $f^+ = f \vee 0$ and $f^- = f \wedge 0$. Hence, $|f| = f^+ - f^-$ is Riemann integrable as well.

Now, we show that $f(\cdot)g(\cdot) \in L_R^\infty(G)$. To this end, we first let

$$f = f^+ - f^-, \qquad g = g^+ - g^-,$$

with

$$\begin{cases} f^+ = f \vee 0, & f^- = -(f \wedge 0), \\ g^+ = g \vee 0, & g^- = -(g \wedge 0). \end{cases}$$

We have the Riemann integrability of $f^\pm$ and $g^\pm$. Further,

$$fg = (f^+ - f^-)(g^+ - g^-) = f^+ g^+ - f^+ g^- - f^- g^+ + f^- g^-.$$

Hence, the Riemann integrability of fg is implied by that of $f^+ g^+$, $f^+ g^-$, $f^- g^+$, and $f^- g^-$. Therefore, it suffices to prove our conclusion for the case that $f, g : G \to [0, M]$, for some $M > 0$.

Now, for any $\alpha \in \mathcal{A}^m$, we define $f_\pm^\alpha(\cdot)$ and $g_\pm^\alpha(\cdot)$ as in (1.24). Then (comparing with (2.7))

$$\begin{aligned} 0 \leqslant f_-^\alpha(x) \leqslant f(x) \leqslant f_+^\alpha(x), \\ 0 \leqslant g_-^\alpha(x) \leqslant g(x) \leqslant g_+^\alpha(x), \end{aligned} \qquad x \in G^\alpha. \tag{2.14}$$

This implies (since all the involved terms are non-negative)

$$f_-^\alpha(x)g_-^\alpha(x) \leqslant f(x)g(x) \leqslant f_+^\alpha(x)g_+^\alpha(x), \quad x \in G^\alpha. \tag{2.15}$$

By the Riemann integrability of $f(\cdot)$ and $g(\cdot)$, for any $\varepsilon > 0$, there exists a $\beta \in \mathcal{A}^m$ such that (2.11) holds. Also, by (2.12),

$$\begin{aligned} 0 &\leqslant f_+^\alpha(x)g_+^\alpha(x) - f_-^\alpha(x)g_-^\alpha(x) \\ &= f_+^\alpha(x)\big[g_+^\alpha(x) - g_-^\alpha(x)\big] + g_-^\alpha(x)\big[f_+^\alpha(x) - f_-^\alpha(x)\big] \\ &\leqslant M\big[g_+^\alpha(x) - g_-^\alpha(x)\big] + M\big[f_+^\alpha(x) - f_-^\alpha(x)\big] = Mh^\alpha(x). \end{aligned}$$

Hence, using (2.13),

$$\begin{aligned} 0 \leqslant S_+(fg) - S_-(fg) &\leqslant \int_{G^\alpha} \big[f_+^\alpha(x)g_+^\alpha(x) - f_-^\alpha(x)g_-^\alpha(x)\big]dx \\ &\leqslant M \int_{G^\alpha} h^\alpha(x)dx \leqslant 4M\varepsilon. \end{aligned}$$

Since $\varepsilon > 0$ is arbitrary, we obtain the Riemann integrability of $f(\cdot)g(\cdot)$.

(iv) By (iii), we have that $f(\cdot)\mathbf{1}_{G_1}(\cdot)$ and $f(\cdot)\mathbf{1}_{G_2}(\cdot)$ are absolutely Riemann integrable, and

$$f(x) = f(x)\mathbf{1}_{G_1}(x) + f(x)\mathbf{1}_{G_2}(x), \qquad x \in G.$$

Then (i) applies to get (2.6). $\qquad\square$

Note that the above, (iii) implies that $L_R^\infty(G) \subseteq L_R^1(G)$. Combining with (1.50), we have

$$L_R^\infty(G) \subseteq L_R^1(G) \subsetneqq L_R(G). \tag{2.16}$$

Now, we briefly look at vector-valued function case $f : G \to \mathbb{R}^\ell$. We have the following definition.

Definition 2.2. Let $f : G \to \mathbb{R}^\ell$, $f(\cdot) \equiv (f^1(\cdot), \cdots, f^\ell(\cdot))$. We say that $f(\cdot)$ is Riemann integrable on G if each component $f^i(\cdot) \in L_R(G)$. In this case, we denote

$$\int_G f(x)dx = \left(\int_G f^1(x)dx, \cdots, \int_G f^\ell(x)dx \right)^\top \in \mathbb{R}^\ell.$$

We call it the *definite Riemann integral* of function $f(\cdot)$ on G. The set of all Riemann integrable $\mathbb{R}^\ell$-valued functions is denoted by $L_R(G; \mathbb{R}^\ell)$. Similar to scalar functions, we denote $L_R^p(G; \mathbb{R}^\ell)$ to be the set of all functions $f(\cdot) \in L_R(G; \mathbb{R}^\ell)$ such that $|f(\cdot)|^p \in L_R(G)$.

It is possible to use the upper and lower Riemann sums to define the Riemann integral of vector-valued functions. We leave the details to the interested readers. Note that the definite integral of an $\mathbb{R}^\ell$-valued function $f(\cdot)$ is a vector in $\mathbb{R}^\ell$. For Riemann integrable vector-valued functions, we have the following result.

Proposition 2.3. *Let $G \subseteq \mathbb{R}^m$ be a bounded domain satisfying* (H1).

(i) *Both $L_R(G; \mathbb{R}^\ell)$ and $L_R^p(G; \mathbb{R}^\ell)$ are linear spaces (for all $p \geqslant 1$).*

(ii) *For any $f(\cdot) \in L_R^1(G; \mathbb{R}^\ell)$,*

$$\left\| \int_G f(x)dx \right\| \leqslant \int_G \|f(x)\| dx. \tag{2.17}$$

(iii) *Let $\bar{G} = \bar{G}_1 \bigcup \bar{G}_2$ with both G_1 and G_2 satisfying* (H1), *and* (2.5). *Let $f(\cdot) \in L_R^1(G; \mathbb{R}^\ell)$. Then $f(\cdot) \in L_R^1(G_1; \mathbb{R}^\ell)$, $f(\cdot) \in L_R^1(G_2; \mathbb{R}^\ell)$, and*

$$\int_G f(x)dx = \int_{G_1} f(x)dx + \int_{G_2} f(x)dx. \tag{2.18}$$

Proof. The proofs of (i) and (iii) are straightforward. We now look at (ii). By definition,

$$\left\| \int_G f(x)dx \right\|^2 = \sum_{i=1}^\ell \left| \int_G f^i(x)dx \right|^2 = \sum_{i=1}^\ell \int_G f^i(x) \left(\int_G f^i(y)dy \right) dx$$

$$= \int_G \sum_{i=1}^\ell f^i(x) \left(\int_G f^i(y)dy \right) dx = \int_G \left[f(x) \cdot \int_G f(y)dy \right] dx$$

$$\leqslant \int_G \|f(x)\| \left\| \int_G f(y)dy \right\| dx = \left\| \int_G f(y)dy \right\| \int_G \|f(x)\| dx.$$

Thus, (2.17) follows.   □

Exercises

1. Suppose $f : [a, b] \to [0, \infty)$ is continuous. Then

$$\int_a^b f dx = 0 \quad \Longleftrightarrow \quad f(x) = 0, \quad \forall x \in [a, b].$$

What if f is piecewise continuous?

2. Let $f : G \to \mathbb{R}$ be a uniformly continuous function on a bounded domain G. Then $f(\cdot)$ is Riemann integrable on G. (Note: G is open.)

3. Let $f(x) = \sin \frac{1}{x}$ which is defined on $(0, 1)$. Is it uniformly continuous on $(0, 1)$? Is it Riemann integrable on $(0, 1)$? Why? What conclusion that you can draw from your conclusions of the above?

4. Let $f : (a, b) \to \mathbb{R}$ be a bounded monotone function. Then $f(\cdot)$ is Riemann integrable.

5*. Let $f, g : [a, b] \to \mathbb{R}$ be continuous functions. Then

$$\int_a^b |f(x)g(x)| dx \leqslant \Big(\int_a^b |f(x)|^p dx \Big)^{\frac{1}{p}} \Big(\int_a^b |g(x)|^q dx \Big)^{\frac{1}{q}},$$

where $p, q \in (1, \infty)$, $\frac{1}{p} + \frac{1}{q} = 1$.

6*. Let $f : [a, b] \to \mathbb{R}^\ell$ be continuous and $p \in (1, \infty)$. Then

$$\Big\| \int_a^b f(x) dx \Big\|_p \leqslant \int_a^b \|f(x)\|_p dx.$$

3   Further Theorems

In this section, we present some more results on Riemann integrals. The first result is the *absolute continuity* of Riemann integral, which is the property of the Riemann integral stated in the following proposition.

Proposition 3.1. *Let $G \subseteq \mathbb{R}^m$ satisfy (1.47), and $f(\cdot) \in L_R(G)$. Then for any $\varepsilon > 0$, there exists a $\delta > 0$ such that for any countable family $\{G_\alpha,\ \alpha \in A\}$ of disjoint subdomains of G, provided*

$$\Big| \bigcup_{\alpha \in A} G_\alpha \Big| < \delta, \qquad |\partial G_\alpha| = 0, \quad \forall \alpha \in A, \tag{3.1}$$

the following holds

$$\Big| \sum_{\alpha \in A} \int_{G_\alpha} f(x) dx \Big| < \varepsilon. \tag{3.2}$$

Proof. First of all, if $f(\cdot)$ is bounded by M, the result is trivial. In fact, we need only to take $\delta = \frac{\varepsilon}{M}$. Now, for general case, we write $f = f^+ - f^-$ with $f^+ = f \vee 0$, $f^- = (-f) \vee 0$. By definition, for any $x_0 \in \mathbb{R}^m$ and $\varepsilon > 0$, there exist $M, N_2 > 0$ such that (recall $G_M(x_0) = G \cap B(x_0, M)$)

$$0 \leqslant \int_G f^+(x)dx - \int_{G_M(x_0)} [f^+]_0^{N_2}(x)dx$$

$$= \int_{G \backslash G_M(x_0)} f^+(x)dx + \int_{G_M(x_0)} \left(f^+(x) - [f^+]_0^{N_2}(x) \right)dx < \frac{\varepsilon}{4}.$$

Let $\delta = \frac{\varepsilon}{4N_2}$. Then for any countable family $\{G_\alpha,\ \alpha \in A\}$ of disjoint subdomains of G satisfying (3.1), denote $G_0 = \cup_{\alpha \in A} G_\alpha$. We have

$$\int_{G_0} f^+(x)dx = \int_{G_0} \left(f^+(x) - [f^+]_0^{N_2}(x) \right)dx + \int_{G_0} [f^+]_0^{N_2}(x)dx$$

$$\leqslant \int_{G_0 \backslash G_M(x_0)} \left(f^+(x) - [f^+]_0^{N_2}(x) \right)dx$$

$$+ \int_{G_0 \cap G_M(x_0)} \left(f^+(x) - [f^+]_0^{N_2}(x) \right)dx + \frac{\varepsilon}{4}$$

$$\leqslant \int_{G \backslash G_M(x_0)} f^+(x)dx + \int_{G_M(x_0)} \left(f^+(x) - [f^+]_0^{N_2}(x) \right)dx + \frac{\varepsilon}{4} < \frac{\varepsilon}{2}.$$

We can do the same thing for $f^-(\cdot)$. Then combining them, we obtain our conclusion. □

Note that in $\mathbb{R}^m$, any family of disjoint non-empty open sets must be countable. (Why?)

The following is a one-dimensional case.

Corollary 3.2. *Let* $f : [a, b] \to \mathbb{R}$ *be Riemann integrable. Then for any* $\varepsilon > 0$, *there exists a* $\delta > 0$, *such that for any family of disjoint intervals* $(a_i, b_i) \subseteq [a, b]$, $i \geqslant 1$, *as long as*

$$\sum_{i \geqslant 1} (b_i - a_i) < \delta,$$

the following holds:

$$\left| \sum_{i \geqslant 1} \int_{a_i}^{b_i} f(x)dx \right| < \varepsilon.$$

Next result is called the *Fundamental Theorem of Calculus* which consists of two parts, the differential form and the integral form.

Theorem 3.3. *Let* $f : [a, b] \to \mathbb{R}$ *be Riemann integrable.*

(i) *Let* $F : [a, b] \to \mathbb{R}$ *be defined by*

$$F(x) = \int_a^x f(s)ds. \tag{3.3}$$

Then $F(\cdot)$ *is continuous on* $[a, b]$. *Further, if* $f(\cdot)$ *is continuous at* $x_0 \in (a, b)$, *then* $F(\cdot)$ *has derivative* $F'(x_0)$ *at* x_0 *and*

$$F'(x_0) = f(x_0). \tag{3.4}$$

(ii) *Let* $f(\cdot)$ *have at most countably many discontinuous points and* $F : [a, b] \to \mathbb{R}$ *be an anti-derivative* $f(\cdot)$, *i.e., at every continuous point* x_0 *of* $f(\cdot)$,

$$F'(x_0) = f(x_0).$$

Then the following Newton-Leibniz formula holds:

$$\int_a^b f(x)dx = F(b) - F(a). \tag{3.5}$$

Proof. (i) The continuity of $F(\cdot)$ follows from Corollary 3.2. Now, let $x_0 \in (a, b)$ at which $f(\cdot)$ is continuous. Then for any $\varepsilon > 0$, there exists a $\delta > 0$ such that

$$|f(x) - f(x_0)| < \varepsilon, \qquad \forall |x - x_0| < \delta, \ x \in [a, b].$$

Next, for any $x \in [a, b]$ with $|x - x_0| < \delta$, one has

$$|F(x) - F(x_0) - f(x_0)(x - x_0)| = \left| \int_{x_0}^x f(s)ds - f(x_0)(x - x_0) \right|$$
$$= \left| \int_{x_0}^x \Big(f(s) - f(x_0) \Big)ds \right| \leqslant \int_{x_0}^x |f(s) - f(x_0)|ds < \varepsilon|x - x_0|.$$

Hence, $F(\cdot)$ has derivative $F'(x_0)$ at x_0 and (3.4) holds.

(ii) We only prove the case that $f(\cdot)$ is bounded and has finitely many discontinuous points listed as follows:

$$a = a_0 < a_1 < \cdots < a_{k-1} < a_k < a_{k+1} = b.$$

By mean-value theorem, we have

$$F(a_i) - F(a_{i-1}) = F'(\xi_i)(a_i - a_{i-1}) = f(\xi_i)(a_i - a_{i-1}),$$
$$i = 1, 2, \cdots, k + 1,$$

for some $\xi_i \in (a_{i-1}, a_i)$. Hence,

$$
S_-^\alpha(f) = \sum_{i=1}^{k+1} \inf_{x \in [a_{i-1}, a_i)} f(x)(a_i - a_{i-1}) \leqslant \sum_{i=1}^{k+1} f(\xi_i)(a_i - a_{i-1})
$$

$$
= \sum_{i=1}^{k+1} \left[F(a_i) - F(a_{i-1}) \right] = F(b) - F(a)
$$

$$
\leqslant \sum_{i=1}^{k+1} \sup_{x \in [a_{i-1}, a_i)} f(x)(a_i - a_{i-1}) = S_+^\alpha(f),
$$

where $\alpha = \{a_0, a_1, \cdots, a_k, a_{k+1}\} \in \mathcal{A}^1$. Also,

$$
S_-^\alpha(f) \leqslant \int_a^b f(x)dx \leqslant S_+^\alpha(f).
$$

Thus,

$$
\left| F(b) - F(a) - \int_a^b f(x)dx \right| \leqslant S_+^\alpha(f) - S_-^\alpha(f).
$$

Note that the left-hand side is independent of α. Hence, letting $\|\alpha\| \to 0$, we obtain our conclusion.

The general case is left to the readers. $\square$

The following result is called the *integration by parts*.

Proposition 3.4. *Let* $F, G : [a, b] \to \mathbb{R}$ *have derivatives such that* $F'(\cdot)$ *and* $G'(\cdot)$ *are Riemann integrable on* $[a, b]$. *Then*

$$
\int_a^b F(x)G'(x)dx = F(b)G(b) - F(a)G(a) - \int_a^b F(x)'G(x)dx.
$$

Proof. By the Newton–Leibniz formula, we have

$$
F(b)G(b) - F(a)G(a) = \int_a^b \left[F(x)G(x) \right]' dx
$$

$$
= \int_a^b F'(x)G(x)dx + \int_a^b F(x)G'(x)dx.
$$

Then our conclusion follows. $\square$

Riemann integral has two mean-value theorems. We now present the *First Mean-Value Theorem*.

Theorem 3.5. *Let* $G \subseteq \mathbb{R}^m$ *be a bounded domain satisfying* (H1). *Let* $f, g : \bar{G} \to \mathbb{R}$ *be continuous, with* $g(x) \geqslant 0$ *for all* $x \in \bar{G}$. *Then there exists a* $\xi \in \bar{G}$ *such that*

$$
\int_G f(x)g(x)dx = f(\xi) \int_G g(x)dx. \tag{3.6}
$$

Proof. Since $g(\cdot)$ is continuous and non-negative, thus if

$$\int_G g(x)dx = 0,$$

we must have

$$g(x) = 0, \qquad x \in \bar{G}.$$

For such a case, (3.6) is obviously true. We now assume that

$$\int_G g(x)dx > 0.$$

By the continuity of $f(\cdot)$ on the compact set $\bar{G}$, we may let

$$\lambda = \min_{x \in \bar{G}} f(x), \qquad \Lambda = \max_{x \in \bar{G}} f(x).$$

Then

$$\lambda g(x) \leqslant f(x)g(x) \leqslant \Lambda g(x), \qquad x \in \bar{G}.$$

Consequently, by Proposition 2.1, (ii), one has

$$\lambda \int_G g(x)dx \leqslant \int_G f(x)g(x)dx \leqslant \Lambda \int_G g(x)dx.$$

This implies that

$$\lambda \leqslant \frac{\displaystyle\int_G f(x)g(x)dx}{\displaystyle\int_G g(x)dx} \leqslant \Lambda.$$

Now, by the intermediate value theorem (Corollary 2.8 of Chapter 2), we can find a $\xi \in \bar{G}$ such that

$$f(\xi) = \frac{\displaystyle\int_G f(x)g(x)dx}{\displaystyle\int_G g(x)dx},$$

which is equivalent to (3.6). $\qquad\qquad\qquad\qquad\qquad\qquad\qquad\square$

Let us look at a couple of special cases of the above.

(i) If $g(x) \equiv 1$, then (3.6) reads

$$\int_G f(x)dx = f(\xi)|G|. \tag{3.7}$$

In particular, when $G = [a, b]$, one has

$$\int_a^b f(x)dx = f(\xi)(b - a).\tag{3.8}$$

(ii) If $G = [a, b]$, then (3.6) reads

$$\int_a^b f(x)g(x)dx = f(\xi)\int_a^b g(x)dx.\tag{3.9}$$

Next, we present the *Second Mean-Value Theorem* of Riemann integrals.

Theorem 3.6. *Let $g : [a, b] \to \mathbb{R}$ be continuous and $f : [a, b] \to \mathbb{R}$ be C^1 such that*

$$f'(x) \geqslant 0, \quad \forall x \in (a, b).$$

Then there exists a $\xi \in (a, b)$ such that

$$\int_a^b f(x)g(x)dx = f(a)\int_a^\xi g(x)dx + f(b)\int_\xi^b g(x)dx.\tag{3.10}$$

Proof. Let

$$G(x) = \int_a^x g(s)ds, \quad x \in [a, b].$$

Using integration by parts, together with (3.6), one has (note $G(a) = 0$)

$$\int_a^b f(x)g(x)dx = \int_a^b f(x)G'(x)dx = f(b)G(b) - \int_a^b f'(x)G(x)dx$$

$$= f(b)G(b) - G(\xi)\int_a^b f'(x)dx = f(b)[G(b) - G(\xi)] + f(a)G(\xi)$$

$$= f(a)\int_a^\xi g(x)dx + f(b)\int_\xi^b g(x)dx,$$

proving our conclusion. $\square$

The following result is called *Fubini's Theorem* which says that, under proper conditions, multiple integral is equal to iterated integrals, and the order of iterated integrals can be exchanged.

Theorem 3.7. (Fubini) *Let $G = G_1 \times G_2$ with $G_1 \subseteq \mathbb{R}^{m_1}$ and $G_2 \subseteq \mathbb{R}^{m_2}$ being bounded domains satisfying*

$$|\partial G_1| = 0, \quad in \ \mathbb{R}^{m_1}; \quad |\partial G_2| = 0, \quad in \ \mathbb{R}^{m_2}.$$

Let $f : \bar{G} \to \mathbb{R}$ be continuous. Then

$$\int_G f(x)dx = \int_{G_1} \left[\int_{G_2} f(x_1, x_2)dx_2 \right] dx_1$$
$$= \int_{G_2} \left[\int_{G_1} f(x_1, x_2)dx_1 \right] dx_2,$$

(3.11)

where the right-hand sides are called *iterated integrals*.

Proof. First, since $f(\cdot)$ is continuous on $\bar{G} = \bar{G}_1 \times \bar{G}_2$ which is compact, it is uniformly continuous. Therefore, the following integrals are well-defined and the following defined maps are continuous:

$$x_1 \mapsto \int_{G_1} f(x_1, x_2)dx_2, \qquad x_2 \mapsto \int_{G_2} f(x_1, x_2)dx_1.$$

Hence, we have the existence of the iterated integrals:

$$\int_{G_1} \left[\int_{G_2} f(x_1, x_2)dx_2 \right] dx_1, \qquad \int_{G_2} \left[\int_{G_1} f(x_1, x_2)dx_1 \right] dx_2.$$

Now, we show (3.11). For any $\alpha_1 \in \mathcal{A}^{m_1}$ and $\alpha_2 \in \mathcal{A}^{m_2}$, let

$$\alpha = (\alpha_1, \alpha_2) \equiv \left\{ (\alpha_{11}^j, \cdots, \alpha_{1m_1}^j, \alpha_{21}^j, \cdots, \alpha_{2m_2}^j) \mid \right.$$
$$\left. (\alpha_{i1}^j, \cdots \alpha_{im_i}^j) \in \alpha_i, \ i = 1, 2 \right\}.$$

Next, similar to the discussion at the beginning of Section 1, we let $\mathcal{Q}_{\alpha_i}(G_i)$ be the smallest subsets of $\mathcal{Q}_{\alpha_i}$ covering $\bar{G}_i$, and let

$$G_{i,\ell}^{\alpha_i} = \bar{G}_i \cap Q_\ell^{\alpha_i}, \qquad 1 \leqslant \ell \leqslant \bar{\ell}_{\alpha_i}, \quad i = 1, 2,$$
$$G_{k\ell}^{\alpha} = G_{1,k}^{\alpha_1} \times G_{2,\ell}^{\alpha_2}, \qquad 1 \leqslant k \leqslant \bar{\ell}_{\alpha_1}, \quad 1 \leqslant \ell \leqslant \bar{\ell}_{\alpha_2}.$$

Then

$$\int_{G_1} \left[\int_{G_2} f(x_1, x_2)dx_2 \right] dx_1 = \lim_{\|\alpha_1\| \to 0} \sum_{k \geqslant 1} \left[\int_{G_2} f(x_{1,k}^*, x_2)dx_2 \right] |G_{1,k}^{\alpha_1}|$$
$$= \lim_{\|\alpha_1\| \to 0} \lim_{\|\alpha_2\| \to 0} \sum_{k,\ell} f(x_{1,k}^*, x_{2,\ell}^*)|G_{1,k}^{\alpha_1}| |G_{2,\ell}^{\alpha_2}|$$
$$= \lim_{\|\alpha\| \to 0} \sum_{k,\ell} f(x_{k\ell}^*)|G_{k\ell}| = \int_G f(x)dx.$$

This proves a half of (3.11). The other half can be proved similarly. $\quad\square$

By some small modifications, we could relax the conditions assumed in the above theorem. We leave the details to the readers. See problem 7 below.

Exercises

1. Let $f : [0, 1] \to \mathbb{R}$ be continuous such that

$$f(1) = 0, \quad \int_0^1 x f(x)^2 dx = 1.$$

Then

$$\int_0^1 x^2 f(x) f'(x) dx = -2,$$

and

$$\left(\int_0^1 |f'(x)|^2 dx \right) \left(\int_0^1 x^4 f(x)^2 dx \right) \geq 4.$$

2. Let $f : [a, b] \to \mathbb{R}$ be differentiable. Then

$$f(b) - f(a) = \left(\int_0^1 f'(a + t(b - a)) dt \right) (b - a).$$

3. Let $g : [0, 1] \to (0, \infty)$ be continuous. Suppose the following holds:

$$a_n \int_0^1 x^n g(x) dx + a_{n-1} \int_0^1 x^{n-1} g(x) dx + \cdots$$

$$+ a_1 \int_0^1 x g(x) dx + a_0 \int_0^1 g(x) dx = 0.$$

Then the equation $a_n x^n + a_{n-1} x^{n-1} + \cdots + a_1 x + a_0 = 0$ has a solution in $[0, 1]$.

4. Let $g : [a, b] \to \mathbb{R}$ be continuous and $f : [a, b] \to \mathbb{R}$ be C^1. Further, $f(\cdot)$ has a unique maximum at $\eta \in (a, b)$. Then there are points $\xi_1 \in (a, \eta)$ and $\xi_2 \in (\eta, b)$ such that

$$\int_a^b f(x) g(x) dx = f(a) \int_a^{\xi_1} g(x) dx + f(\eta) \int_{\xi_1}^{\xi_2} g(x) dx + f(b) \int_{\xi_2}^b g(x) dx.$$

5. Let $f : [0, \infty) \to \mathbb{R}$ such that

$$\lim_{x \to \infty} f(x) = m,$$

for some $m \in \mathbb{R}$. Let $\ell > 0$ fixed. Find $\lim_{x \to \infty} \int_x^{x+\ell} f(s) ds$.

6. Let $G_k \subseteq \mathbb{R}^{m_k}$ be bounded domains, $k = 1, 2$ such that

$$|\partial G_2| = 0, \quad \text{in } \mathbb{R}^{m_2}.$$

Let $f : \bar{G}_1 \times \bar{G}_2 \to \mathbb{R}$ be continuous. Define

$$F(x_1) = \int_{G_2} f(x_1, x_2) dx_2, \qquad x_1 \in \bar{G}_1.$$

Then $F : \bar{G}_1 \to \mathbb{R}$ is continuous.

7. Let $G_1 \subseteq \mathbb{R}^{m_1}$ and $G_2 \subseteq \mathbb{R}^{m_2}$ be bounded sets with

$$|\partial G_1| = 0, \quad |\partial G_2| = 0,$$

in $\mathbb{R}^{m_1}$ and in $\mathbb{R}^{m_2}$, respectively. Let $f : G_1 \times G_2 \to \mathbb{R}$ be bounded and let

$$\begin{aligned}
D(f) &= \big\{(x_1, x_2) \in G_1 \times G_2 \mid f(\cdot) \text{ is discontinuous at } (x_1, x_2)\big\}, \\
D\big(f(x_1, \cdot)\big) &= \big\{x_2 \in G_2 \mid f(x_1, \cdot) \text{ is discontinuous at } x_2\big\}, \\
D\big(f(\cdot, x_2)\big) &= \big\{x_1 \in G_2 \mid f(\cdot, x_2) \text{ is discontinuous at } x_1\big\}.
\end{aligned}$$

Suppose

$$\begin{aligned}
|D(f)| &= 0, &&\text{in } \mathbb{R}^{m_1 + m_2}, \\
\big|D\big(f(x_1, \cdot)\big)\big| &= 0, &&\text{in } \mathbb{R}^{m_2}, \quad \forall x_1 \in G_1, \\
\big|D\big(f(\cdot, x_2)\big)\big| &= 0, &&\text{in } \mathbb{R}^{m_1}, \quad \forall x_2 \in G_2.
\end{aligned}$$

Then Fubini Theorem holds.

8. Let $G \subseteq \mathbb{R}^m$ be a bounded domain with $|\partial G| = 0$ in $\mathbb{R}^m$, let $f, g : \bar{G} \to \mathbb{R}$ be continuous. Let $p \in (1, \infty)$. Then

$$\Big(\int_G |f(x) + g(x)|^p dx\Big)^{\frac{1}{p}} \leqslant \Big(\int_G |f(x)|^p dx\Big)^{\frac{1}{p}} + \Big(\int_G |g(x)|^p dx\Big)^{\frac{1}{p}}.$$

9. Let $G_1 \subseteq \mathbb{R}^{m_1}$ and $G_2 \subseteq \mathbb{R}^{m_2}$ be bounded sets with

$$|\partial G_1| = 0, \quad |\partial G_2| = 0,$$

in $\mathbb{R}^{m_1}$ and in $\mathbb{R}^{m_2}$, respectively. Let $f : \bar{G}_1 \times \bar{G}_2 \to \mathbb{R}$ be continuous. Then for any $p \in [1, \infty)$,

$$\Big(\int_{G_1} \Big|\int_{G_2} f(x_1, x_2) dx_2\Big|^p dx_1\Big)^{\frac{1}{p}} \leqslant \int_{G_2} \Big(\int_{G_1} |f(x_1, x_2)|^p dx_1\Big)^{\frac{1}{p}} dx_2.$$

Chapter 5

Uniform Convergence

1 Observations and Examples

Let us begin with some observations.

- *Observation 1.* Suppose $\{a_{nm}\}_{n,m \geqslant 1}$ is a sequence with double indices. We may expect or hope that

$$\lim_{n \to \infty} \lim_{m \to \infty} a_{nm} = \lim_{m \to \infty} \lim_{n \to \infty} a_{nm}. \tag{1.1}$$

- *Observation 2.* Suppose $\{f_n(\cdot)\}_{n \geqslant 1}$ is a sequence of continuous functions defined on $[a, b]$, $n \geqslant 1$, such that

$$\lim_{n \to \infty} f_n(x) = f(x), \qquad x \in [a, b].$$

We may expect or hope that $f(\cdot)$ is continuous. In other words, for any $x_0 \in [a, b]$, we hope

$$\lim_{x \to x_0} \lim_{n \to \infty} f_n(x) = \lim_{x \to x_0} f(x) = f(x_0) = \lim_{n \to \infty} \lim_{x \to x_0} f_n(x). \tag{1.2}$$

- *Observation 3.* Suppose $\{f_n(\cdot)\}_{n \geqslant 1}$ is a sequence of differentiable functions defined on $[a, b]$ and converges to some function $f(\cdot)$. Then we may expect or hope that $f(\cdot)$ is differentiable and

$$f'(x) = \lim_{n \to \infty} f_n'(x). \tag{1.3}$$

This is equivalent to the following:

$$\lim_{h \to 0} \lim_{n \to \infty} \frac{f_n(x+h) - f_n(x)}{h} = \lim_{n \to \infty} \lim_{h \to 0} \frac{f_n(x+h) - f_n(x)}{h}. \tag{1.4}$$

- *Observation 4.* Suppose $\{f_n(\cdot)\}_{n \geqslant 1}$ is a sequence of Riemann integrable functions on $[a, b]$ and $f_n(\cdot)$ is convergent to $f(\cdot)$. We may expect or hope that $f(\cdot)$ is Riemann integrable and

$$\int_a^b f(x)dx = \lim_{n \to \infty} \int_a^b f_n(x)dx. \tag{1.5}$$

This is equivalent to the following:

$$\lim_{m \to \infty} \lim_{n \to \infty} \sum_{i=1}^{m} f_n(x_i^m) \Delta x_i^m = \lim_{n \to \infty} \lim_{m \to \infty} \sum_{i=1}^{m} f_n(x_i^m) \Delta x_i^m, \qquad (1.6)$$

where $\{x_i^m\}_{0 \leqslant i \leqslant m}$ is a partition of $[a, b]$ (labeled by m), and $\Delta x_i^m = x_i^m - x_{i-1}^m$.

Let us look at the following example.

Example 1.1. (i) Let

$$a_{nm} = \frac{n}{n + m}, \qquad n, m \geqslant 1. \qquad (1.7)$$

Then

$$\lim_{n \to \infty} \lim_{m \to \infty} a_{nm} = 0 \neq 1 = \lim_{m \to \infty} \lim_{n \to \infty} a_{nm}.$$

Thus, in general, (1.1) fails.

(ii) Let

$$f_n(x) = \frac{nx}{nx + 1}, \qquad x \in [0, 1], \ n \geqslant 1. \qquad (1.8)$$

Then each $f_n(\cdot)$ is continuous and

$$\lim_{n \to \infty} f_n(x) = \begin{cases} 0, & x = 0, \\ 1, & x \in (0, 1], \end{cases}$$

which is not continuous. Hence, (1.2) is not true in general.

(iii) Let

$$f_n(x) = x - \frac{1}{n} \log(nx + 1), \qquad x \in [0, 1]. \qquad (1.9)$$

Then

$$f_n'(x) = 1 - \frac{1}{nx + 1} = \frac{nx}{nx + 1},$$

and

$$\lim_{n \to \infty} f_n(x) = x, \qquad x \in [0, 1].$$

But,

$$\left(\lim_{n \to \infty} f_n(x) \right)' = 1 \neq \lim_{n \to \infty} f_n'(x) = \begin{cases} 1, & x \in (0, 1], \\ 0, & x = 0. \end{cases}$$

Therefore, (1.3) is not guaranteed in general.

(iv) Let

$$f_n(x) = n\mathbf{1}_{(0,\frac{1}{n})}(x) \equiv \begin{cases} n, & x \in (0, \dfrac{1}{n}), \\ 0, & x \in [0,1] \setminus (0, \dfrac{1}{n}). \end{cases}$$

Then

$$\int_0^1 f_n(x)dx = 1, \qquad \forall n \geqslant 1,$$

and

$$\lim_{n\to\infty} f_n(x) = 0, \qquad \forall x \in [0,1].$$

Hence,

$$\lim_{n\to\infty} \int_0^1 f_n(x)dx = 1 \neq 0 = \int_0^1 \lim_{n\to\infty} f_n(x)dx.$$

This means that (1.5) is not true in general.

From the above example, we see that all 4 observations listed at the beginning of this section could be false.

2 Uniform Convergence

We now introduce the following definition.

Definition 2.1. Let (X, d_X) and (Y, d_Y) be two metric spaces and $f_n : X \to Y$ be a sequence of functions $(n \geqslant 1)$. Also, let $f : X \to Y$ be another given function.

(i) We say that $f_n(\cdot)$ *pointwisely* converges to $f(\cdot)$ if for any given $x \in X$, the following is true: For any $\varepsilon > 0$, there exists an $N = N(\varepsilon, x) \geqslant 1$ (may depend on x) such that

$$d_Y(f_n(x), f(x)) < \varepsilon, \qquad \forall n \geqslant N.$$

(ii) We say that $f_n(\cdot)$ *uniformly* converges to $f(\cdot)$ if for any $\varepsilon > 0$, there exists an $N = N(\varepsilon) > 0$, (independent of $x \in X$) such that

$$d_Y(f_n(x), f(x)) < \varepsilon, \qquad \forall n \geqslant N, x \in X.$$

(iii) We say that $\{f_n(\cdot)\}_{n\geqslant 1}$ is *uniformly Cauchy* if for any $\varepsilon > 0$, there exists an $N = N(\varepsilon) \geqslant 1$ (independent of x) such that

$$d_Y(f_n(x), f_m(x)) < \varepsilon, \qquad \forall n, m \geqslant N, \ x \in X.$$

Example 2.2. (i) Let $X = [0, 1]$ and $Y = \mathbb{R}$ with the standard metric. Let

$$f_n(x) = \frac{nx}{nx + 1}, \qquad n \geqslant 1.$$

Then as Example 1.1, (ii), one has

$$\lim_{n \to \infty} f_n(x) = \begin{cases} 1, & x \neq 0, \\ 0, & x = 0. \end{cases}$$

We claim that the above convergence is not uniform. In fact, for any $x \neq 0$, in order to have

$$|f_n(x) - 1| = 1 - \frac{nx}{nx + 1} = \frac{1}{nx + 1} < \varepsilon, \quad \forall n \geqslant N,$$

we need

$$Nx > \frac{1}{\varepsilon} - 1 = \frac{1 - \varepsilon}{\varepsilon},$$

which is equivalent to

$$N > \frac{1 - \varepsilon}{\varepsilon x}.$$

Hence, as x is getting smaller and smaller, N has to be getting larger and larger. Therefore, the convergence is not uniform. Now, if instead of $[0, 1]$, we let

$$X = \left\{ \frac{1}{m} \mid m = 1, 2, \cdots \right\} \cup \{0\},$$

with the metric induced from the standard metric of $\mathbb{R}$, then consider the restriction $f_n\big|_X = \tilde{f}_n : X \to Y$, we have

$$\tilde{f}_n\left(\frac{1}{m}\right) = \frac{n}{n + m} = a_{nm}, \qquad f_n(0) = 0,$$

which is the sequence appears in Example 1.1, (i). Hence, the above argument shows that

$$\lim_{n \to \infty} a_{nm} = 1 = \lim_{n \to \infty} f_n\left(\frac{1}{m}\right)$$

is not uniform in m. Consequently, we can show that the convergence

$$\lim_{m \to \infty} a_{nm} = 0$$

is not uniform in n, either.

(ii) Let

$$f_n(x) = x - \frac{1}{n} \log(nx + 1), \qquad x \in [0, 1], \ n \geqslant 1.$$

Then

$$|f_n(x) - x| \leqslant \frac{1}{n} \log(nx + 1) \leqslant \frac{\log(n+1)}{n} \to 0, \qquad n \to \infty,$$

uniformly in $x \in [0,1]$. That is $f_n(x)$ converges to $f(x) = x$ uniformly. However,

$$f_n'(x) = 1 - \frac{1}{nx+1} = \frac{nx}{nx+1}, \qquad x \in [0,1].$$

From part (i), we see that $f_n'(\cdot)$ is not uniformly convergent on $[0,1]$.

   (iii) Let

$$f_n(x) = n\mathbf{1}_{(0,\frac{1}{n})}(x), \qquad x \in [0,1].$$

We see that $f_n(\cdot)$ convergent pointwise on $[0,1]$, but not uniformly. In fact, since the limit function is $f(x) \equiv 0$, in order $f_n(\cdot)$ to converge to $f(\cdot)$, it is necessary that $f_n(\cdot)$ is uniformly bounded. But the above $f_n(\cdot)$ is not uniformly bounded.

The following result gives us some more feeling about the pointwise convergence and the uniform convergence.

Proposition 2.3. *Let* $f : \mathbb{R} \to \mathbb{R}$, *and for any* $a \in \mathbb{R}$, *define the shift function as follows:*

$$f^a(x) = f(x - a), \qquad x \in \mathbb{R}.$$

Then $f(\cdot)$ *is (uniformly) continuous on* $\mathbb{R}$ *if and only if for any sequence* a_n *converging to 0, the sequence* $f^{a_n}(\cdot)$ *of shift functions is (uniformly) convergent to* $f(\cdot)$.

Proof. Suppose $f(\cdot)$ is continuous. Then for any $a_n \to 0$,

$$\lim_{n \to \infty} f^{a_n}(x) = \lim_{n \to \infty} f(x - a_n) = f(x).$$

Conversely, if $f(\cdot)$ is discontinuous at $\bar{x}$, then there is a sequences a_n convergent to 0, such that for some $\varepsilon > 0$,

$$\varepsilon \leqslant |f(\bar{x} - a_n) - f(\bar{x})| = |f^{a_n}(\bar{x}) - f(\bar{x})|.$$

This means that $f^{a_n}(\cdot)$ is not convergent to $f(\cdot)$, a contradiction.

Next, we look at the uniform continuity case. First, suppose $f(\cdot)$ is uniformly continuous. Let a_n be an arbitrary sequence that is convergent to 0. By the uniform continuity of $f(\cdot)$, for any $\varepsilon > 0$, there exists a $\delta = \delta(\varepsilon) > 0$, independent of x, such that

$$|f(x) - f(\bar{x})| < \varepsilon, \qquad \forall x, \bar{x} \in \mathbb{R}, |x - \bar{x}| < \delta.$$

Now since $a_n \to 0$, there exists an $N \geqslant 1$, depending on $\delta > 0$ such that

$$|a_n| < \delta, \qquad \forall n \geqslant N.$$

Hence, for any $x \in \mathbb{R}$, by taking $\bar{x} = x - a_n$, one has

$$|f^{a_n}(x) - f(x)| = |f(\bar{x}) - f(x)| < \varepsilon,$$

proving the uniform convergence of $f^{a_n}(\cdot)$.

Conversely, suppose $f(\cdot)$ is not uniformly continuous. Then there exist two sequences x_n, a_n such that for some $\varepsilon_0 > 0$,

$$|a_n| \to 0, \qquad |f(x_n - a_n) - f(x_n)| \geqslant \varepsilon_0, \quad n \geqslant 1.$$

Then, it leads to that $f^{a_n}(\cdot)$ is not uniformly convergent to $f(\cdot)$, a contradiction. $\qquad\square$

We now present the following result which will be used below.

Theorem 2.4. *Let (X, d_X) and (Y, d_Y) be metric spaces, and $E \subseteq X$, $f_n, f : E \to Y$, $n \geqslant 1$. Suppose $f_n(\cdot)$ converges to $f(\cdot)$ uniformly. Then $\{f_n(\cdot)\}_{n \geqslant 1}$ is uniformly Cauchy. Conversely, if $\{f_n(\cdot)\}_{n \geqslant 1}$ is uniformly Cauchy and (Y, d_Y) is complete, then $\{f_n(\cdot)\}_{n \geqslant 1}$ is uniformly convergent.*

Proof. Suppose $f_n(\cdot)$ is uniformly convergent to $f(\cdot)$. Then for any $\varepsilon > 0$, there exists an $N = N(\varepsilon) \geqslant 1$ such that

$$d_Y\big(f_n(x), f(x)\big) < \varepsilon, \qquad \forall n \geqslant N, \ x \in X.$$

Consequently, for any $n, m \geqslant N$, one has

$$d_Y\big(f_n(x), f_m(x)\big) \leqslant d_Y\big(f_n(x), f(x)\big) + d_Y\big(f_m(x), f(x)\big) < 2\varepsilon, \qquad \forall x \in X.$$

Thus, $\{f_n(\cdot)\}_{n \geqslant 1}$ is uniformly Cauchy.

Conversely, let $\{f_n(\cdot)\}_{n \geqslant 1}$ be uniformly Cauchy. Hence, for any $\varepsilon > 0$, there exists an $N \geqslant 1$ such that

$$d_Y(f_n(x), f_m(x)) < \varepsilon, \qquad \forall n, m \geqslant N, \ \forall x \in X.$$

Then for each $x \in X$, $\{f_n(x)\}_{n \geqslant 1}$ is a Cauchy sequence in Y. Since (Y, d_Y) is complete, we may define

$$f(x) = \lim_{n \to \infty} f_n(x), \qquad x \in X.$$

Consequently, for any $x \in X$, we can find an $\widetilde{N} = \widetilde{N}(\varepsilon, x) \geqslant N$ such that

$$d_Y\big(f_n(x), f(x)\big) < \varepsilon, \qquad \forall n \geqslant \widetilde{N}(\varepsilon, x).$$

This leads to that for any $n \geqslant N$,

$$d_Y\big(f_n(x), f(x)\big) \leqslant d_Y\big(f_n(x), f_{\widetilde{N}(\varepsilon,x)}(x)\big) + d_Y\big(f_{\widetilde{N}(\varepsilon,x)}(x), f(x)\big) < 2\varepsilon.$$

This shows that $f_n(\cdot)$ converges to $f(\cdot)$ uniformly. $\qquad\square$

The following is a main result of this section.

Theorem 2.5. *Let (X, d_X) and (Y, d_Y) be metric spaces, and $E \subseteq X$, $f_n, f : E \to Y$, $n \geqslant 1$. Suppose $f_n(\cdot)$ converges to $f(\cdot)$ uniformly. Then the following statements are true.*

(i) *Suppose x_0 is a limit point of E, and*

$$\lim_{E \ni x \to x_0} f_n(x) = L_n, \qquad n \geqslant 1,$$

for some $L_n \in Y$. Then, $\{L_n\}_{n \geqslant 1}$ is Cauchy in Y. In the case that (Y, d_Y) is complete, there exists some $L \in Y$ such that

$$\lim_{n \to \infty} L_n = L = \lim_{E \ni x \to x_0} f(x),$$

which amounts to saying that

$$\lim_{n \to \infty} \lim_{E \ni x \to x_0} f_n(x) = \lim_{E \ni x \to x_0} \lim_{n \to \infty} f_n(x).$$

(ii) *Suppose for each $n \geqslant 1$, $f_n(\cdot)$ is continuous. Then $f(\cdot)$ is continuous.*

Proof. (i) By Theorem 2.4, for any $\varepsilon > 0$, there exists an $N = N(\varepsilon) \geqslant 1$ such that

$$d_Y\big(f_n(x), f_m(x)\big) < \varepsilon, \qquad \forall n, m \geqslant N, \ x \in E.$$

Also, for x_0, a limit point of E, for any $n \geqslant 1$, there exists a $\delta_n = \delta(\varepsilon, x_0, n) > 0$ such that

$$d_Y(f_n(x), L_n) \leqslant \varepsilon, \qquad \forall x \in E \cap B_X(x_0, \delta_n).$$

Then for any $n, m \geqslant N$, we choose $x_{nm} \in B_X(x_0, \delta_n \wedge \delta_m) \cap E$ to get the following:

$$d_Y(L_n, L_m) \leqslant d_Y\big(L_n, f_n(x_{nm})\big) + d_Y\big(f_n(x_{nm}), f_m(x_{nm})\big)$$
$$+ d_Y\big(f_m(x_{nm}), L_m\big) < 3\varepsilon,$$

which shows that $\{L_n\}_{n \geqslant 1}$ is Cauchy. Consequently, if (Y, d_Y) is complete, we have some $L \in Y$ so that

$$\lim_{n \to \infty} d_Y(L_n, L) = 0.$$

Then for any $\varepsilon > 0$, there exists an $\widetilde{N} \geqslant N$ such that

$$d_Y(L_n, L) < \varepsilon, \qquad \forall n \geqslant \widetilde{N}.$$

Hence, for any $x \in E \cap B_X(x_0, \delta_{\widetilde{N}})$, we have

$$d_Y(f(x), L) \leqslant d_Y(f(x), f_{\widetilde{N}}(x)) + d_Y(f_{\widetilde{N}}(x), L_{\widetilde{N}}) + d_Y(L_{\widetilde{N}}, L) < 3\varepsilon,$$

which proves our claim.

(ii) It follows from (i) with $L_n = f_n(x_0)$ and $L = f(x_0)$. $\qquad \square$

Uniform convergence of a continuous function sequence is a sufficient condition for the limit function to be continuous, and it is not necessary. Here is a simple example.

Example 2.6. Let

$$f_n(x) = \begin{cases} nx, & x \in [0, n^{-1}], \\ 2 - nx, & x \in [n^{-1}, 2n^{-1}], \\ 0, & x \in [2n^{-1}, 1]. \end{cases}$$

Then $f_n(\cdot)$ is continuous and $f_n \to 0$ pointwise, but not uniform. Whereas, the limit function $f(\cdot)$ is continuous.

Thus, a natural question is what kind of convergence for the sequence of continuous functions will exactly lead to the continuity of the limit function? We now introduce the following notion.

Definition 2.7. Let (X, d) be a metric space and $f_n, f : X \to \mathbb{R}$. We say that $f_n(\cdot)$ is *quasi-uniformly converges* to $f(\cdot)$ if $f_n(\cdot) \to f(\cdot)$ pointwise on X, and for any $\varepsilon > 0$, $N \geqslant 1$, there are at most countably many open sets $\{G_i\}_{i \geqslant 1}$, and corresponding indices $\{n_i\}_{i \geqslant 1}$ such that

$$X \subseteq \bigcup_{i \geqslant 1} G_i,$$

and

$$|f_{n_i}(x) - f(x)| < \varepsilon, \qquad \forall x \in G_i, \quad i \geqslant 1.$$

We have the following interesting result.

Theorem 2.8. *Let (X, d) be a compact metric space and $f_n : X \to \mathbb{R}$ be a sequence of continuous functions convergent pointwise to some $f : X \to \mathbb{R}$. Then $f_n(\cdot)$ converges to $f(\cdot)$ quasi-uniformly if and only if f is continuous.*

Proof. Sufficiency. For any $\varepsilon > 0$ and $N \geqslant 1$, by the uniform continuity of $f(\cdot)$ on the compact metric space X, there exists a $\delta_\varepsilon > 0$ such that

$$|f(x) - f(y)| < \varepsilon, \qquad \forall x, y \in X, \ d(x, y) < \delta_\varepsilon.$$

By the pointwise convergence of $f_n(\cdot)$ to $f(\cdot)$, for each $x \in X$, there exists an $N_x \geqslant N$ such that

$$|f_{N_x}(x) - f(x)| < \varepsilon.$$

Further, by the continuity of $f_{N_x}(\cdot)$, there exists an $\alpha_x \in (0, \delta_\varepsilon)$ such that

$$|f_{N_x}(y) - f_{N_x}(x)| < \varepsilon, \qquad \forall y \in X, \ d(x, y) < \alpha_x.$$

Hence, for any $y \in X$, $d(x, y) < \alpha_x$,

$$\begin{aligned}
|f(y) - f_{N_x}(y)| &\leqslant |f(y) - f(x)| + |f(x) - f_{N_x}(x)| \\
&\quad + |f_{N_x}(x) - f_{N_x}(y)| < 3\varepsilon.
\end{aligned}$$

Clearly, the family $\{B_{\alpha_x}(x) \mid x \in X\}$ is an open cover of X. By the compactness of X, we may assume that

$$X \subseteq \bigcup_{i=1}^{k} B_{\alpha_{x_i}}(x_i) \equiv \bigcup_{i=1}^{k} G_i.$$

Also, let $n_i = N_{x_i}$. Then

$$|f_{n_i}(y) - f(y)| < 3\varepsilon, \qquad \forall y \in X \cap G_i, \ 1 \leqslant i \leqslant k.$$

This means that $f_n(\cdot) \to f(\cdot)$ quasi-uniformly.

Necessity. Fix any $x \in X$. For any $\varepsilon > 0$, by the definition of quasi-uniform convergence, there exists an open set G containing x and an N such that

$$|f_N(x) - f(x)| < \varepsilon,$$

and

$$|f_N(y) - f(y)| < \varepsilon, \qquad \forall y \in G.$$

By the continuity of f_N, there exists a $\delta > 0$ such that

$$|f_N(x) - f_N(y)| < \varepsilon, \qquad \forall y \in G, \ d(x, y) < \delta.$$

Then for any $y \in G$, $d(y, x) < \delta$,

$$|f(x) - f(y)| \leqslant |f(x) - f_N(x)| + |f_N(x) - f_N(y)| + |f_N(y) - f(y)| < 3\varepsilon.$$

Thus, $f(\cdot)$ is continuous at x. $\qquad\square$

We have seen that uniform limit of continuous function sequence on a compact set is continuous (Theorem 2.5, (ii)); the pointwise limit of continuous function sequence on a compact set is continuous if and only if the convergence is quasi-uniform (Theorem 2.8). Now, we would like to know when a pointwise convergent continuous function sequence is convergent uniformly. Here is such a result.

Theorem 2.9. (Dini) *Suppose (X, d) is a compact metric space and $f_n : X \to \mathbb{R}$ is a sequence of continuous functions. Suppose*

$$\lim_{n \to \infty} f_n(x) = f(x), \qquad x \in X,$$

with $f : X \to \mathbb{R}$ being continuous. Further, suppose that

$$f_n(x) \leqslant f_{n+1}(x), \qquad x \in X. \tag{2.1}$$

Then the convergence of $f_n(\cdot)$ to $f(\cdot)$ is uniform.

Proof. Let

$$g_n(x) = f(x) - f_n(x), \qquad x \in G. \tag{2.2}$$

Then each $g_n(\cdot)$ is continuous,

$$g_n(x) \geqslant g_{n+1}(x) \geqslant 0, \qquad \forall x \in X, \ n \geqslant 1, \tag{2.3}$$

and

$$\lim_{n \to \infty} g_n(x) = 0. \tag{2.4}$$

Now, for any $\varepsilon > 0$, let

$$\mathcal{O}_n(\varepsilon) = g_n^{-1}((-\infty, \varepsilon)) \equiv \{x \in X \mid g_n(x) < \varepsilon\}. \tag{2.5}$$

Then by the continuity of $g_n(\cdot)$, $\mathcal{O}_n(\varepsilon)$ is open. Moreover, by (2.3), we have

$$\mathcal{O}_n(\varepsilon) \subseteq \mathcal{O}_{n+1}(\varepsilon), \qquad \forall n \geqslant 1. \tag{2.6}$$

On the other hand, by (2.4), for each $x \in X$, there exists an N such that

$$g_n(x) < \varepsilon, \qquad \forall n \geqslant N, \tag{2.7}$$

which means that $x \in \mathcal{O}_n(\varepsilon)$. Hence,

$$X \subseteq \bigcup_{n=1}^{\infty} \mathcal{O}_n(\varepsilon). \tag{2.8}$$

Now, by the compactness of X, we have a finite subset of $\{\mathcal{O}_n(\varepsilon) \mid n \geqslant 1\}$ covering X. Let N be the largest index of this finite set. Then by (2.6), we must have

$$\mathcal{O}_N(\varepsilon) \supseteq X.$$

This means that

$$0 \leqslant f(x) - f_n(x) \equiv g_n(x) \leqslant g_N(x) < \varepsilon, \qquad \forall n \geqslant N, \ x \in X.$$

Hence, the convergence of $f_n(\cdot)$ to $f(\cdot)$ is uniform. $\qquad\square$

It is clear that condition (2.1) can be replaced by

$$f_n(x) \geqslant f_{n+1}(x), \qquad x \in X.$$

We also note that in Dini's Theorem, X can be any compact metric space, in particular, it could be any bounded and closed set in $\mathbb{R}^m$, and it does not have to be the closure of some domain.

Exercises

1. Let (X, d) be a metric space and $f_n : X \to \mathbb{R}$ be a continuous function for each $n \geqslant 1$ such that $f_n(\cdot) \to f(\cdot)$ uniformly for some $f : X \to \mathbb{R}$. Suppose $x_n \in E$ such that $x_n \to \bar{x} \in E$. Then

$$\lim_{n\to\infty} f_n(x_n) = f(\bar{x}).$$

2. Let (Y, d) be a metric space and $a_{nm} \in Y$, $n, m \geqslant 1$, such that

$$\lim_{n\to\infty} a_{nm} = b_m, \qquad \text{uniformly in } m \geqslant 1,$$

and

$$\lim_{m\to\infty} a_{nm} = c_n.$$

Then c_n is Cauchy in Y and in the case that Y is complete, there exists a $\bar{c}$ such that

$$\lim_{n\to\infty}\lim_{m\to\infty} a_{nm} = \lim_{n\to\infty} c_n = \bar{c} = \lim_{m\to\infty} b_m = \lim_{m\to\infty}\lim_{n\to\infty} a_{nm}.$$

3. Let (X, d) be a metric space and $f_n, g_n : X \to \mathbb{R}$ such that $f_n \to f$ and $g_n \to g$ uniformly, for some $f, g : X \to \mathbb{R}$.

(i) Show that for any $\alpha, \beta \in \mathbb{R}$, $\alpha f_n + \beta g_n \to \alpha f + \beta g$ uniformly.

(ii) Prove or disprove that if $\alpha_n, \beta_n \in \mathbb{R}$ with $\alpha_n \to \alpha$ and $\beta_n \to \beta$ for some $\alpha, \beta \in \mathbb{R}$, then $\alpha_n f_n + \beta_n g_n \to \alpha f + \beta g$ uniformly.

(iii) Show that in general "$f_n g_n \to fg$ uniformly" is not true. But it is true if both f_n and g_n are uniformly bounded.

4. Let (X, d_X) and (Y, d_Y) be metric spaces and $f_n : X \to Y$ be uniformly convergent to some $f : X \to Y$. Let $g : Y \to \mathbb{R}$ be Lipschitz continuous. Prove or disprove that $g \circ f_n$ converges to $g \circ f$ uniformly.

5. Let

$$f_n(s) = \begin{cases} 0, & 0 < x < \dfrac{1}{n+1}, \text{ or } x > \dfrac{1}{n}, \\[2mm] \sin^2 \dfrac{\pi}{x}, & \dfrac{1}{n+1} \leqslant x \leqslant \dfrac{1}{n}. \end{cases}$$

Show that $f_n(\cdot)$ converges to a continuous function, but the convergence is not uniform.

6. Let $f : [0, 1] \to \mathbb{R}$ be a continuous function. Show that there exists a sequence of step functions $f_k : [0, 1] \to \mathbb{R}$ such that $f_k(\cdot)$ converges to $f(\cdot)$ uniformly. Is the result true if the domain $[0, 1]$ is replaced by $(0, 1)$? Why?

7. Let (X, d) be a metric space which is totally bounded, not necessarily complete, and $f_n : X \to \mathbb{R}$ be a sequence of uniformly continuous functions. Suppose

$$f_n(x) \leqslant f_{n+1}(x), \qquad x \in X, \ n \geqslant 1,$$

and

$$\lim_{n \to \infty} f_n(x) = f(x), \qquad x \in X,$$

with some uniformly continuous function $f : X \to \mathbb{R}$. Show that the convergence is uniform. What if each $f_n(\cdot)$ is just continuous? Give your reason.

8. Let (X, d) be a compact metric space and $f_n : X \to \mathbb{R}$ converge to $f : X \to \mathbb{R}$ uniformly. Then f_n converges to f quasi-uniformly. Is the converse true? Why?

3 The Metric of Uniform Convergence

We now look at the uniform convergence of functions from a different viewpoint.

Proposition 3.1. *Let (X, d_X) and (Y, d_Y) be complete metric spaces, let*

$$B(X; Y) = \big\{ f : X \to Y \mid f(\cdot) \text{ is bounded} \big\},$$
$$C(X; Y) = \big\{ f(\cdot) \in B(X; Y) \mid f(\cdot) \text{ is continuous} \big\},$$

and define

$$d_\infty(f,g) = \sup_{x\in X} d_Y\big(f(x),g(x)\big), \qquad \forall f(\cdot), g(\cdot) \in B(X;Y).$$

(i) d_∞ is a metric on $B(X;Y)$, which is called the *sup norm metric* or *L^∞-metric*. A sequence $\{f_n\}_{n\geqslant 1}$ in $B(X;Y)$ converges to $f(\cdot) \in B(X;Y)$ in the metric d_∞ if and only if $f_n(\cdot)$ converges to $f(\cdot)$ uniformly.

(ii) The set $C(X;Y)$ is complete as a subspace of $B(X;Y)$, i.e., any Cauchy sequence $\{f_n(\cdot)\}_{n\geqslant 1}$ in $C(X;Y)$ is convergent to some $f(\cdot) \in C(X;Y)$ under d_∞.

Proof. The proof of (i) is straightforward.

(ii) Let $\{f_n(\cdot)\}_{n\geqslant 1}$ be a Cauchy sequence in $C(X;Y)$. Then by (i), $f_n(\cdot)$ converges to some $f(\cdot) \in B(X;Y)$. By Theorem 2.5 (ii), $f(\cdot) \in C(X;Y)$. $\quad\square$

We now introduce a new notion.

Definition 3.2. Let (X,d_X) and (Y,d_Y) be metric spaces. A family $\mathcal{F} \subseteq C(X;Y)$ is said to be *equi-continuous* at $x_0 \in X$ if for any $\varepsilon > 0$, there exists a $\delta = \delta(x_0, \varepsilon) > 0$, independent of the particular $f(\cdot) \in \mathcal{F}$, such that

$$d_Y\big(f(x), f(x_0)\big) < \varepsilon, \qquad \forall d_X(x, x_0) < \delta, \ f(\cdot) \in \mathcal{F}.$$

The family $\mathcal{F}$ is said be *uniformly equi-continuous* if for any $\varepsilon > 0$, there exists a $\delta = \delta(\varepsilon) > 0$ such that

$$d_Y\big(f(x), f(\bar{x})\big) < \varepsilon, \qquad \forall d_X(x, \bar{x}) < \delta, \ f(\cdot) \in \mathcal{F}.$$

The following result collects some interesting relations among some relevant notions.

Theorem 3.3. Let (X,d_X) be a compact metric space and (Y,d_Y) be a complete metric space. Let $f_n : X \to Y$ be continuous such that $f_n(\cdot)$ converges to $f(\cdot)$ pointwise on X. Then the following are equivalent:

(i) The sequence $\{f_n(\cdot)\}$ is equicontinuous on X.

(ii) Function $f(\cdot)$ is continuous on X and the convergence $f_n(\cdot) \to f(\cdot)$ is uniform on X.

Proof. (i) $\Rightarrow$ (ii): Since (X,d_X) is compact, the equi-continuity of $\{f_n(\cdot)\}$ is uniform in $x \in X$. That is, for any $\varepsilon > 0$, there exists a $\delta > 0$ (only depending on $\varepsilon > 0$) such that the following holds:

$$d_Y\big(f_n(x), f_n(x')\big) < \varepsilon, \qquad \forall x, x' \in X, \ d_X(x, x') < \delta, \ n \geqslant 1.$$

Letting $n \to \infty$, we obtain

$$d_Y\big(f(x), f(x_0)\big) \leqslant \varepsilon. \tag{3.1}$$

Thus, $f(\cdot)$ is continuous.

Next, by the compactness of X, for the above $\delta > 0$, there are finitely many $x_1, \cdots, x_k$ such that

$$X \subseteq \bigcup_{i=1}^{k} B(x_i, \delta),$$

and

$$d_Y\big(f_n(x), f_n(x_i)\big) < \varepsilon, \qquad \forall x \in B(x_i, \delta), \quad 1 \leqslant i \leqslant k, \qquad \forall n \geqslant 1. \tag{3.2}$$

Further, by the pointwise convergence $f_n(\cdot) \to f(\cdot)$, for the above $\varepsilon > 0$, there exists an $N \geqslant 1$ such that

$$d_Y\big(f_n(x_i), f(x_i)\big) < \varepsilon, \qquad \forall n \geqslant N, \; 1 \leqslant i \leqslant k. \tag{3.3}$$

Consequently, for any $y \in X$, letting $y \in B(x_i, \delta)$ for some $1 \leqslant i \leqslant k$, we have (noting (3.1), (3.2), and (3.3))

$$\begin{aligned}
d_Y\big(f_n(y), f(y)\big) &\leqslant d_Y\big(f_n(y), f_n(x_i)\big) + d_Y\big(f_n(x_i), f(x_i)\big) \\
&\quad + d_Y\big(f(x_i), f(y)\big) < 3\varepsilon, \qquad \forall n \geqslant N.
\end{aligned}$$

This means that the convergence $f_n(\cdot) \to f(\cdot)$ is uniform.

(ii) $\Rightarrow$ (i): Let $x_0 \in X$. For any $\varepsilon > 0$, there exists a $\delta > 0$ such that

$$d_Y\big(f(x), f(x_0)\big) < \varepsilon, \qquad \forall x \in B(x_0, \delta),$$

and there exists an $N \geqslant 1$ (independent of $x \in X$) such that

$$d_Y\big(f_n(x), f(x)\big) < \varepsilon, \qquad \forall n \geqslant N, \; x \in X.$$

Then for any $n \geqslant N$, and $x \in B(x_0, \delta)$,

$$\begin{aligned}
d_Y\big(f_n(x), f_n(x_0)\big) &\leqslant d_Y\big(f_n(x), f(x)\big) + d_Y\big(f(x), f(x_0)\big) \\
&\quad + d_Y\big(f(x_0), f_n(x_0)\big) < 3\varepsilon.
\end{aligned}$$

Also, for each $n = 0, 1, \cdots, N - 1$, there exists a $\delta_n > 0$ such that

$$d_Y\big(f_n(x), f_n(x_0)\big) < \varepsilon, \qquad x \in B(x_0, \delta).$$

Then by taking

$$\bar{\delta} = \min\{\delta, \delta_0, \delta_1, \cdots, \delta_{N-1}\},$$

we see that for any $x \in B(x_0, \delta)$,

$$d_Y\big(f_n(x), f_n(x_0)\big) < 3\varepsilon,$$

proving the equicontinuity of $\{f_n(\cdot)\}$ near x_0. Then by the compactness of X, we obtain (i). $\qquad\square$

We see that in the above, (ii) $\Rightarrow$ (i) does not need the compactness of X.

The following result can be regarded as an extension of Bolzano–Weierstrass Theorem to a family of functions, which gives a criterion for compact subsets in $C(X; Y)$.

Theorem 3.4. (Arzelà–Ascoli) *Let (X, d_X) and (Y, d_Y) be compact metric spaces. Let $\mathcal{F} = \{f_\lambda : X \to Y, \lambda \in \Lambda\}$ be a family of functions. Suppose that $\mathcal{F}$ is equicontinuous. Then there exists a sequence $\{f_{\lambda_n}(\cdot)\} \subseteq \mathcal{F}$ which converges to some continuous function $f : X \to Y$ uniformly on X.*

Proof. Since X is compact, by Proposition 4.18 of Chapter 1, there exists a countable set $X_0 \equiv \{x_n, n \geqslant 1\} \subseteq X$ which is dense in X, i.e.,

$$\bar{X}_0 = X.$$

Now, for $x_1 \in X_0$, since $\{f_\lambda(x_1) \mid \lambda \in \Lambda\}$ is a sequence in the compact metric space Y, there exists a subsequence $\{f_{\lambda_k}(x_1)\} \subseteq Y$ that is convergent to some point denoted by $f(x_1) \in Y$. Next, since $\{f_{\lambda_k}(x_2)\} \subseteq Y$ is a sequence in the compact metric space Y, we can further extract a subsequence of $\{f_{\lambda_k}(x_2)\}$ which is convergent to some point $f(x_2) \in Y$. By relabeling, we may assume that $\{f_{\lambda_k}(x_2)\}$ itself is convergent to some point denoted by $f(x_2)$. Continue this procedure, we obtain a sequence $\{f_{\lambda_k}(\cdot)\} \subseteq \mathcal{F}$ such that

$$\lim_{k \to \infty} f_{\lambda_k}(x_n) = f(x_n), \qquad \forall x_n \in X_0, \tag{3.4}$$

where $f(x_n)$ is some point in Y depending on x_n. Next, we claim that $f_{\lambda_k}(\cdot)$ is uniformly convergent on X. In fact, for any $\varepsilon > 0$, by the equicontinuity of $\mathcal{F}$, we have a $\delta > 0$, independent of $k \geqslant 1$, such that

$$d_Y\big(f_{\lambda_k}(x), f_{\lambda_k}(\bar{x})\big) < \varepsilon, \qquad x, \bar{x} \in X, \ d_X(x, \bar{x}) < \delta, \ k \geqslant 1.$$

By the compactness of X, there are points $x_1, \cdots, x_n \in X_0$, with n depending on $\delta > 0$, such that

$$X \subseteq \bigcup_{i=1}^{n} B(x_i, \delta).$$

By (3.4), there exists an $N \geqslant 1$, depending on $\varepsilon > 0$ (and $n \geqslant 1$, which eventually depends on $\varepsilon > 0$), such that

$$d_Y\big(f_{\lambda_j}(x_i), f_{\lambda_k}(x_i)\big) < \varepsilon, \qquad j, k \geqslant N, \ 1 \leqslant i \leqslant n. \qquad (3.5)$$

Now, for any $x \in X$, let $x_i \in X_0$ such that

$$d_X(x_i, x) < \delta.$$

Then for $j, k \geqslant N$,

$$\begin{aligned}
d_Y\big(f_{\lambda_j}(x), f_{\lambda_k}(x)\big) &\leqslant d_Y\big(f_{\lambda_j}(x), f_{\lambda_j}(x_i)\big) + d_Y\big(f_{\lambda_j}(x_i), f_{\lambda_k}(x_i)\big) \\
&\quad + d_Y\big(f_{\lambda_k}(x_i), f_{\lambda_k}(x)\big) < 3\varepsilon.
\end{aligned}$$

This shows that $\{f_{\lambda_k}(\cdot)\}$ is Cauchy in $\big(C(X;Y), d_\infty\big)$. Hence, we may denote

$$f(x) = \lim_{k \to \infty} f_{\lambda_k}(x), \qquad \text{uniformly in } x \in X.$$

Then, by Theorem 3.3, it is necessary that $f(\cdot)$ is continuous. $\qquad\square$

A well-known form of Arzelà–Ascoli Theorem is the following corollary.

Corollary 3.5. *Let $G \subseteq \mathbb{R}^n$ be a compact set, and $f_n : G \to \mathbb{R}^m$ be a sequence of functions. Then $\{f_n(\cdot)\}_{n \geqslant 1}$ has a subsequence which is uniformly convergent if the sequence is uniformly bounded and equicontinuous.*

Next result gives an approximation of continuous functions by polynomials.

Theorem 3.6. (Weierstrass) *If $f : [a, b] \to \mathbb{R}$ is continuous. Then there exists a sequence of polynomials $P_n(\cdot)$ such that*

$$\lim_{n \to \infty} \|P_n(\cdot) - f(\cdot)\|_\infty \equiv \sup_{x \in [a,b]} |P_n(x) - f(x)| = 0.$$

Proof. First, without loss of generality, we assume that $[a, b] = [0, 1]$. Further, we assume

$$f(0) = f(1) = 0.$$

In fact, if necessary, instead of $f(\cdot)$ on $[0, 1]$, we may consider

$$g(x) = f(x) - f(0) - x[f(1) - f(0)], \qquad x \in [0, 1].$$

If such a $g(\cdot)$ can be approximated by polynomials $p_n(\cdot)$ uniformly, i.e.,

$$\lim_{n \to \infty} \|g(\cdot) - p_n(\cdot)\|_\infty = 0.$$

Then

$$\lim_{n\to\infty} \|f(\cdot) - f(0) - x[f(1) - f(0)] - p_n(\cdot)\|_\infty = 0,$$

and $p_n(\cdot) + f(0) + x[f(1) - f(0)]$ is a polynomial.

Next, we redefine $f(\cdot)$ to be zero outside of $[0,1]$. Then $f(\cdot)$ is bounded and uniformly continuous on $\mathbb{R}$. Let

$$M = \sup_{x\in[0,1]} |f(x)|.$$

Define

$$Q_n(x) = c_n(1 - x^2)^n, \qquad n \geqslant 1,$$

with c_n being chosen so that

$$\int_{-1}^{1} Q_n(x)dx = 1.$$

Consider

$$\varphi(x) = (1 - x^2)^n - 1 + nx^2, \qquad x \in [0, 1].$$

We have $\varphi(0) = 0$, and

$$\varphi'(x) = -n(1 - x^2)^{n-1}2x + 2nx = 2nx[1 - (1 - x^2)^{n-1}] \geqslant 0.$$

Hence, $\varphi(x) \geqslant 0$ for all $x \in [0, 1]$, which implies

$$(1 - x^2)^n \geqslant 1 - nx^2, \qquad x \in [0, 1].$$

Consequently,

$$\frac{1}{c_n} = \int_{-1}^{1}(1 - x^2)^n dx = 2\int_{0}^{1}(1 - x^2)^n dx \geqslant 2\int_{0}^{\frac{1}{\sqrt{n}}}(1 - x^2)^n dx$$

$$\geqslant 2\int_{0}^{\frac{1}{\sqrt{n}}}(1 - nx^2)dx = 2\left(x - \frac{n}{3}x^3\right)\Big|_{0}^{\frac{1}{\sqrt{n}}} = \frac{4}{3\sqrt{n}} > \frac{1}{\sqrt{n}}.$$

This yields

$$c_n = \left(\int_{-1}^{1}(1 - x^2)^n dx\right)^{-1} < \sqrt{n}.$$

Thus, for any given $\delta \in (0, 1)$,

$$Q_n(x) \leqslant \sqrt{n}(1 - x^2)^n \leqslant \sqrt{n}(1 - \delta^2)^n, \qquad \delta \leqslant x \leqslant 1,$$

which implies

$$\lim_{n\to\infty} Q_n(x) = 0, \qquad \text{uniformly in } x \in [\delta, 1].$$

Note that $f(x) = 0$ for $x \notin [0,1]$. Thus, we have

$$P_n(x) \equiv \int_{-1}^{1} f(x+t)Q_n(t)dt = \int_{-x}^{1-x} f(x+t)Q_n(t)dt$$

$$= \int_{0}^{1} f(t)Q_n(t-x)dt, \qquad x \in (0,1].$$

Clearly, $P_n(\cdot)$ is a polynomial. Now, for any $\varepsilon > 0$, there exists a $\delta > 0$ such that

$$|f(x) - f(\bar{x})| < \frac{\varepsilon}{2}, \qquad \forall x, \bar{x} \in [0,1], \ |x - \bar{x}| < \delta.$$

Then for any $x \in [0,1]$,

$$|P_n(x) - f(x)| = \left| \int_{-1}^{1} [f(x+t) - f(x)]Q_n(t)dt \right|$$

$$\leqslant \int_{-1}^{1} |f(x+t) - f(x)|Q_n(t)dt$$

$$\leqslant 2M \left(\int_{-1}^{-\delta} Q_n(t)dt + \int_{\delta}^{1} Q_n(t)dt \right) + \frac{\varepsilon}{2} \int_{-\delta}^{\delta} Q_n(t)dt$$

$$= 4M \int_{\delta}^{1} Q_n(t)dt + \frac{\varepsilon}{2} \int_{-1}^{1} Q_n(t)dt$$

$$\leqslant 4M\sqrt{n}(1-\delta^2)^n + \frac{\varepsilon}{2} < \varepsilon,$$

as long as n large enough. This proves the theorem. □

The following gives an application of the above theorem.

Example 3.7. Let $f : [a,b] \to \mathbb{R}$ be continuous such that for any polynomial $p(\cdot)$,

$$\int_{a}^{b} f(x)p(x)dx = 0.$$

Then $f(\cdot) = 0$. To show this, we choose a sequence of polynomials $p_n(\cdot)$ such that

$$\|f(\cdot) - p_n(\cdot)\|_{\infty} < \frac{1}{n}, \qquad \forall n \geqslant 1.$$

By our condition, we have

$$\int_{a}^{b} |f(x)|^2 dx = \int_{a}^{b} f(x)[f(x) - p_n(x)]dx$$

$$\leqslant \int_{a}^{b} |f(x)||f(x) - p_n(x)|dx \leqslant \frac{M}{n}(b-a) \to 0.$$

Thus, $f(\cdot) = 0$.

A generalization of the above Weierstrass Theorem is the following.

Theorem 3.8. (Müntz–Szász) *Let* $0 = \lambda_0 < \lambda_1 < \cdots$. *In* $C[0,1]$, *consider the following subset:*

$$\Pi(\Lambda) = \text{span}\{x^{\lambda_k} \mid k = 0, 1, 2, \cdots\}.$$

Then $\Pi(\Lambda)$ *is dense in* $C[0,1]$ *if and only if*

$$\sum_{k=1}^{\infty} \frac{1}{\lambda_k} = \infty.$$

The proof of this theorem is beyond the scope of this book and we omit it here. Weierstrass Theorem is a special case of the above with $\lambda_k = k$.

Exercises

1. Let $f_n : [0,1] \to \mathbb{R}$ be a sequence of continuously differentiable functions such that

$$f_n(0) = 0, \quad |f_n'(x)| \leqslant 1, \quad \forall n \geqslant 1, \ x \in (0,1).$$

Suppose further that $f_n(\cdot)$ is convergent pointwise to some function $f(\cdot)$. Then $f_n(\cdot)$ converges to $f(\cdot)$ uniformly.

2. Let $f_n(x) = x^n$, $x \in [0,1]$. Show that for any $\delta \in (0,1)$, the sequence f_n is equicontinuous on $[0,\delta]$, but the sequence is not equicontinuous on $[0,1]$.

3. Let $f_n : [a,b] \to \mathbb{R}$ be continuous and uniformly bounded. Define

$$g_n(x) = \int_a^x f_n(s)ds, \quad x \in [a,b], \ n \geqslant 1.$$

Then $g_n(\cdot)$ is equicontinuous.

4. Let $f : [0,1] \to \mathbb{R}$ be continuous such that

$$\int_0^1 x^k f(x)dx = 0, \quad \forall k \geqslant 5.$$

Then $f(x) = 0$ for all $x \in [0,1]$.

5. Let $f : [0,1] \to \mathbb{R}$ be continuous such that

$$\int_0^1 x^{2k} f(x)dx = 0, \quad \forall k \geqslant 1.$$

Then $f(x) = 0$, for all $x \in [0,1]$.

6. Let $K : [0,1] \times [0,1] \to \mathbb{R}$ be continuous. Let $f_n : [0,1] \to \mathbb{R}^n$ be continuous and uniformly bounded. Define

$$g_n(x) = \int_0^1 K(x,y)f_n(y)dy, \quad x \in [0,1], \ n \geqslant 1.$$

Then the sequence $\{g_n(\cdot)\}$ is uniformly bounded and equicontinuous.

4   Limit Theorems

In this section, we consider the order exchanges. The first result is the order exchange of the limit and integration.

Theorem 4.1. *Let $G \subseteq \mathbb{R}^m$ be a bounded domain with $|\partial G| = 0$, and $f_n : G \to \mathbb{R}$ be a sequence of Riemann integrable functions. Suppose that $f_n(\cdot)$ uniformly converges to $f(\cdot)$ on G. Then $f(\cdot)$ is Riemann integrable and*

$$\lim_{n \to \infty} \int_G f_n(x)dx = \int_G f(x)dx \equiv \int_G \lim_{n \to \infty} f_n(x)dx.$$

Proof. Since $f_n(\cdot)$ converges to $f(\cdot)$ uniformly, for any $\varepsilon > 0$, there exists an $N \geqslant 1$ such that

$$|f_n(x) - f(x)| < \varepsilon, \qquad \forall n \geqslant N, \quad x \in G.$$

This implies that

$$f_n(x) - \varepsilon < f(x) < f_n(x) + \varepsilon, \qquad \forall n \geqslant N, \quad x \in G.$$

Consequently, for any $\alpha \in \mathcal{A}^m$, (see Chapter 4, (1.26))

$$S_\alpha^-(f_n) - \varepsilon|G^\alpha| \leqslant S_\alpha^-(f) \leqslant S_\alpha^+(f) \leqslant S_\alpha^+(f_n) + \varepsilon|G^\alpha|,$$

which leads to

$$0 \leqslant S_\alpha^+(f) - S_\alpha^-(f) \leqslant S_\alpha^+(f_n) - S_\alpha^-(f_n) + 2\varepsilon|G^\alpha|.$$

Hence,

$$\overline{\lim_{\|\alpha\| \to 0}} \left[S_\alpha^+(f) - S_\alpha^-(f) \right] \leqslant 2\varepsilon|G|.$$

Since $\varepsilon > 0$ is arbitrary, one sees that $f(\cdot)$ is Riemann integrable. The above also implies

$$\left| \int_G f_n(x)dx - \int_G f(x)dx \right| \leqslant 2\varepsilon|G|, \qquad n \geqslant N,$$

proving our conclusion. $\qquad\qquad\qquad\qquad\qquad\qquad\qquad\qquad\qquad\qquad$ $\square$

The following is a consequence of the above theorem and the Dini's Theorem.

Corollary 4.2. (Levi's Lemma) *Let $G \subseteq \mathbb{R}^m$ be a bounded domain with $|\partial G| = 0$. Suppose $f_n : \bar{G} \to \mathbb{R}$ is a sequence of continuous functions such that*

$$\lim_{n \to \infty} f_n(x) = f(x), \qquad x \in \bar{G},$$

with $f : [a, b] \to \mathbb{R}$ being continuous, and

$$f_n(x) \leqslant f_{n+1}(x), \qquad x \in \bar{G}, \quad n \geqslant 1.$$

Then

$$\lim_{n \to \infty} \int_G f_n(x) dx = \int_G f(x) dx.$$

Proof. By Dini's Theorem, the convergence of $f_n(\cdot)$ to $f(\cdot)$ is uniform (since $\bar{G}$ is compact). Hence, Theorem 4.1 applies. $\qquad\square$

The following is another corollary.

Corollary 4.3. (Fatou's Lemma) *Let $G \subseteq \mathbb{R}^m$ be a bounded domain with $|\partial G| = 0$. Suppose $f_n : \bar{G} \to [0, \infty)$ is continuous such that for each $k \geqslant 1$,*

$$g_k(x) = \inf_{n \geqslant k} f_n(x), \qquad x \in \bar{G}$$

is continuous on $\bar{G}$, and

$$\lim_{n \to \infty} f_n(x) = f(x), \qquad x \in \bar{G}$$

is also continuous. Then

$$\int_G \lim_{n \to \infty} f_n(x) dx \leqslant \lim_{n \to \infty} \int_G f_n(x) dx.$$

Proof. Since

$$g_1(x) \leqslant g_k(x) \leqslant g_{k+1}(x), \qquad x \in \bar{G},$$

and

$$\lim_{k \to \infty} g_k(x) = \lim_{n \to \infty} f_n(x) = f(x), \qquad x \in \bar{G}.$$

By Corollary 4.2, we have

$$\lim_{k \to \infty} \int_G g_k(x) dx = \int_G f(x) dx.$$

On the other hand,

$$g_k(x) \leqslant f_k(x), \qquad \forall x \in \bar{G},$$

which leads to

$$\int_G g_k(x) dx \leqslant \int_G f_k(x) dx.$$

Hence,

$$\int_G f(x)dx = \lim_{k \to \infty} \int_G g_k(x)dx \leqslant \lim_{n \to \infty} \int_G f_n(x)dx,$$

proving our result.   □

We know that in Example 1.1, (iv),

$$f_n(x) = n\mathbf{1}_{(0,\frac{1}{n})}(x), \qquad x \in [0,1]$$

which pointwisely converges to 0. But the convergence is not uniform. Therefore, the order exchange of limit and integration is not necessarily true. On the other hand, the uniform convergence of $f_n(\cdot)$ to $f(\cdot)$ is not necessary for the order exchange of limit and integration. Here is an example.

Example 4.4. Let

$$f_n(x) = \mathbf{1}_{(0,\frac{1}{n})}(x), \qquad x \in [0,1].$$

Then

$$\lim_{n \to \infty} f_n(x) = 0, \qquad x \in [0,1],$$

but the convergence is not uniform. However,

$$\int_0^1 f_n(x)dx = \frac{1}{n} \to 0 = \int_0^1 0dx.$$

From the above, we may ask can one relax the condition that ensures the order exchange of the limit and integration? The answer is yes and it can be stated as follows.

Theorem 4.5. (Arzelà) *Let $G \subseteq \mathbb{R}^m$ be a bounded domain with $|\partial G| = 0$. Let $f_n : \bar{G} \to \mathbb{R}$ be Riemann integrable and uniformly bounded. Suppose*

$$\lim_{n \to \infty} f_n(x) = f(x), \qquad x \in \bar{G},$$

for some Riemann integrable function $f : \bar{G} \to \mathbb{R}$. Then

$$\lim_{n \to \infty} \int_G |f_n(x) - f(x)|dx = 0, \tag{4.1}$$

which leads to

$$\lim_{n \to \infty} \int_G f_n(x)dx = \int_G f(x)dx. \tag{4.2}$$

Proof. First, let $f_n : \bar{G} \to \mathbb{R}$ be a decreasing sequence of bounded functions (not necessarily Riemann integrable) such that

$$\lim_{n \to \infty} f_n(x) = 0, \qquad \forall x \in \bar{G}.$$

By the proof of Proposition 1.9 of Chapter 4, for any $\varepsilon > 0$ and any $n \geqslant 1$, there exists a continuous function $g_n^\varepsilon : \bar{G} \to \mathbb{R}$ such that

$$0 \leqslant g_n^\varepsilon(x) \leqslant f_n(x), \qquad \forall x \in \bar{G},$$

and

$$S^-(f_n) \leqslant \int_G g_n^\varepsilon(x)dx + \frac{\varepsilon}{2^n}. \tag{4.3}$$

For each $n \geqslant 1$, set

$$h_n^\varepsilon(x) = \min\left\{g_1^\varepsilon(x), \cdots, g_n^\varepsilon(x)\right\}, \qquad x \in \bar{G}.$$

Then

$$0 \leqslant h_n^\varepsilon(x) \leqslant g_n^\varepsilon(x) \leqslant f_n(x), \qquad x \in \bar{G},$$

and $\{h_n^\varepsilon\}_{n \geqslant 1}$ is decreasingly approaching to 0 on $\bar{G}$ as $n \to \infty$. Hence, by Dini's Theorem, the convergence is uniform. Consequently,

$$\lim_{n \to \infty} \int_G h_n^\varepsilon(x)dx = 0.$$

Now, we claim that

$$0 \leqslant S^-(f_n) \leqslant \int_G h_n^\varepsilon(x)dx + \varepsilon(1 - 2^{-n}), \qquad n \geqslant 1. \tag{4.4}$$

In fact, for each $i = 1, 2, \cdots, n$, one has

$$
\begin{aligned}
0 \leqslant g_n^\varepsilon(x) &= g_i^\varepsilon(x) + g_n^\varepsilon(x) - g_i^\varepsilon(x) \\
&\leqslant g_i^\varepsilon(x) + \max\{g_i^\varepsilon(x), \cdots, g_n^\varepsilon(x)\} - g_i^\varepsilon(x) \\
&\leqslant g_i^\varepsilon(x) + \sum_{j=1}^{n-1} \Big(\max\{g_j^\varepsilon(x), \cdots, g_n^\varepsilon(x)\} - g_j^\varepsilon(x) \Big),
\end{aligned}
$$

which leads to

$$0 \leqslant g_n^\varepsilon(x) \leqslant h_n^\varepsilon(x) + \sum_{j=1}^{n-1} \Big(\max\{g_j^\varepsilon(x), \cdots, g_n^\varepsilon(x)\} - g_j^\varepsilon(x) \Big). \tag{4.5}$$

On the other hand, by

$$\max\{g_j^\varepsilon(x), \cdots, g_n^\varepsilon(x)\} \leqslant \max\{f_j(x), \cdots, f_n(x)\} = f_j(x), \qquad x \in \bar{G},$$

one has

$$S^-(f_j) \geqslant \int_G \Big(\max\{g_j^\varepsilon(x), \cdots, g_n^\varepsilon(x)\} - g_j(x) \Big) dx + \int_G g_j^\varepsilon(x) dx.$$

Thus, by (4.3),

$$\int_G \Big(\max\{g_j^\varepsilon(x), \cdots, g_n^\varepsilon(x)\} - g_j^\varepsilon(x) \Big) dx \tag{4.6}$$

$$\leqslant S^-(f_j) - \int_G g_j^\varepsilon(x) dx < \frac{\varepsilon}{2^j}, \qquad 1 \leqslant j \leqslant n.$$

Consequently, combining (4.5)–(4.6), we obtain

$$\int_G g_n^\varepsilon(x) dx \leqslant \int_G h_n^\varepsilon(x) dx + \sum_{j=1}^{n-1} \frac{\varepsilon}{2^j} = \int_G h_n^\varepsilon(x) dx + \varepsilon\Big(1 - \frac{1}{2^{n-1}}\Big).$$

Hence, (4.4) holds and we obtain

$$\lim_{n \to \infty} S^-(f_n) = 0.$$

Now, we prove our theorem. Without loss of generality, we assume that

$$0 \leqslant f_n(x) \leqslant M, \qquad \forall x \in \bar{G}, \ n \geqslant 1,$$

and $f(x) \equiv 0$ (otherwise, we consider $|f_n(x) - f(x)|$ which is Riemann integrable). We define

$$p_n(x) = \sup_{k \geqslant 0} f_{n+k}(x), \qquad x \in \bar{G}, \quad n \geqslant 1.$$

Note that $p_n(\cdot)$ is not necessarily Riemann integrable. Clearly,

$$0 \leqslant f_n(x) \leqslant p_n(x), \qquad x \in \bar{G}, \ n \geqslant 1,$$

and $\{p_n(\cdot)\}_{n \geqslant 1}$ is a decreasing sequence of bounded functions that converges to 0 pointwise. Hence, by what we have proved,

$$\lim_{n \to \infty} S^-(p_n) = 0.$$

Consequently,

$$0 \leqslant \lim_{n \to \infty} \int_G f_n(x) dx \leqslant \lim_{n \to \infty} S^-(p_n) = 0,$$

proving the theorem. $\qquad\qquad\qquad\qquad\qquad\qquad\qquad\qquad\qquad\qquad\square$

With the above Arzelà's Theorem, we could relax the conditions in the Levi's Lemma and Fatou's Lemma. Details are left to the readers. Note that the condition that $f(\cdot)$ being Riemann integrable cannot be removed as the pointwise limit of a sequence of Riemann integrable functions might be not Riemann integrable. Here is an example.

Example 4.6. Let $\mathbb{Q} \cap [0, 1] = \{r_k \mid k \geqslant 1\}$. Define

$$f_n(x) = \begin{cases} 1, & x \in \{r_k \mid 1 \leqslant k \leqslant n\}, \\ 0, & \text{otherwise.} \end{cases}$$

Then each $f_n(\cdot)$ is Riemann integrable, and

$$\lim_{n \to \infty} f_n(x) = f(x) = \begin{cases} 1, & x \in \mathbb{Q} \cap [0, 1], \\ 0, & x \in [0, 1] \setminus \mathbb{Q}. \end{cases}$$

This is a Dirichlet function and it is not Riemann integrable.

Next, we consider the situation of differentiation. We first look at the following example.

Example 4.7. (i) Let

$$f_n(x) = \frac{\sin nx}{\sqrt{n}}, \qquad x \in [0, 1], \ n \geqslant 1.$$

Then

$$\lim_{n \to \infty} f_n(x) = 0 = f(x), \qquad \text{uniformly in } x \in [0, 1].$$

On the other hand,

$$f_n'(x) = \sqrt{n} \cos nx, \qquad x \in [0, 1],$$

and $f'(x) = 0$. Hence,

$$\left(\lim_{n \to \infty} f_n(x) \right)' = \lim_{n \to \infty} f_n'(x)$$

is not true.

(ii) Let

$$f_n(x) = \sqrt{\frac{1}{n} + x^2}, \qquad x \in [-1, 1], \ n \geqslant 1.$$

Then

$$\lim_{n \to \infty} f_n(x) = |x| = f(x), \qquad x \in [-1, 1],$$

and

$$f_n'(x) = \frac{x}{\sqrt{\frac{1}{n} + x^2}} \to g(x) = \begin{cases} 0, & x = 0, \\ \text{sgn}(x), & x \neq 0. \end{cases}$$

Thus, the limit function $f(\cdot)$ is not differentiable at $x = 0$, and of course

$$\left(\lim_{n \to \infty} f_n(x) \right)' = \lim_{n \to \infty} f_n'(x)$$

is not true.

However, here is a positive result.

Theorem 4.8. *Let $f_n : [a, b] \to \mathbb{R}$ be a sequence of continuously differentiable functions such that*

$$\lim_{n \to \infty} f_n'(x) = g(x), \qquad \text{uniformly in } x \in [a, b],$$

for some $g(\cdot)$, and there exists an $x_0 \in [a, b]$ such that

$$\lim_{n \to \infty} f_n(x_0) = L$$

holds. Then there exists a differentiable function $f(\cdot)$ such that

$$\lim_{n \to \infty} f_n(x) = f(x), \qquad \text{uniformly in } x \in [a, b],$$

and

$$f'(x) = g(x), \qquad x \in [a, b].$$

Proof. We have

$$f_n(x) = f_n(x_0) + \int_{x_0}^x f_n'(t)dt.$$

Now, passing to the limit on the right-hand side, making use of Theorem 4.1, we have

$$f(x) = L + \int_{x_0}^x g(t)dt.$$

Then our conclusion follows. $\qquad\square$

The above amounts to saying that under conditions of Theorem 4.8,

$$\lim_{n \to \infty} f_n'(x) = f'(x) = \left(\lim_{n \to \infty} f_n(x) \right)'.$$

We now look at the multi-variable case.

Proposition 4.9. *Let $f_n : G \to \mathbb{R}$ be a sequence of differentiable functions. Suppose that $\nabla f_n(\cdot)$ is uniformly bounded on G. Then $\{f_n(\cdot)\}$ is equicontinuous.*

Proof. By our assumption, we may let

$$\|\nabla f_n(x)\| \leqslant M, \qquad \forall x \in G. \tag{4.7}$$

Hence, by the mean value theorem, we have

$$\begin{aligned}\|f_n(x) - f_n(y)\| &= \|\nabla f_n(x + \theta(y - x)) \cdot (x - y)\| \\ &\leqslant M\|x - y\|, \qquad \forall x, y \in G.\end{aligned} \tag{4.8}$$

From this, we see immediately that $\{f_n(\cdot)\}$ is equicontinuous on G. $\qquad\square$

Theorem 4.10. *Let $G \subseteq \mathbb{R}^m$ be a bounded domain and $f_n : \bar{G} \to \mathbb{R}$ be a sequence of C^1 functions that converges to a function $f : \bar{G} \to \mathbb{R}$. Moreover, the sequence $\{\nabla f_n(\cdot)\}$ is uniformly bounded and convergent. Then the following are equivalent:*

(i) $\{\nabla f_n(\cdot)\}$ *is equicontinuous on $\bar{G}$.*

(ii) $f(\cdot)$ *is C^1, $\nabla f_n : G \to \mathbb{R}^m$ converges to $\nabla f(\cdot)$ uniformly, and $f_n(\cdot)$ converges to $f(\cdot)$ uniformly as well.*

Proof. (i) $\Rightarrow$ (ii): Since $\{\nabla f_n(\cdot)\}$ is equicontinuous, for any $\varepsilon > 0$, there exists a $\delta > 0$ such that

$$\|\nabla f_n(x) - \nabla f_n(y)\| < \varepsilon, \qquad \forall x, y \in G, \ \|x - y\| < \delta. \qquad (4.9)$$

Now, for a given $x_0 \in G$, suppose $B(x_0, \delta) \subseteq G$ (if necessary, we may shrink $\delta > 0$). By the mean value theorem, we have

$$f_n(x_0 + y) - f_n(x_0) = \nabla f_n(x_0 + \theta_n y) \cdot y, \qquad \theta_n \in (0, 1). \qquad (4.10)$$

Thus,

$$\frac{\|f_n(x_0 + y) - f_n(x_0) - \nabla f_n(x_0) \cdot y\|}{\|y\|}$$
$$\leqslant \|\nabla f_n(x_0 + \theta_n y) - \nabla f_n(x_0)\| < \varepsilon, \qquad \theta_n \in (0, 1), \ \forall \|y\| < \delta.$$

On the other hand, by the uniform boundedness and equi-continuity of $\{\nabla f_n(\cdot)\}$, we may let $\nabla f_{n_k}(\cdot)$ converge to some $g(\cdot)$. Then passing to the limit along this subsequence in the above, we obtain

$$\frac{\|f(x_0 + y) - f(x_0) - g(x_0) \cdot y\|}{\|y\|} \leqslant \varepsilon, \qquad \forall \|y\| < \delta.$$

This implies that $f(\cdot)$ is differentiable at x_0 and

$$\nabla f(x_0) = g(x_0).$$

Since every subsequence of $\{\nabla f_n(x_0)\}$ admits a sub-subsequence converging to the same point $\nabla f(x_0)$, one obtains that the whole sequence $\{\nabla f_n(x_0)\}$ is convergent for all $x_0 \in G$. Further, by the equi-continuity of $\{\nabla f_n(\cdot)\}$ again, we obtain the continuity of $\nabla f(\cdot)$. Then by Theorem 3.3, one has that the convergence $\nabla f_n(\cdot) \to \nabla f(\cdot)$ must be uniform. Since $\nabla f_n(\cdot)$ is uniformly bounded, by Proposition 4.9, $\{f_n(\cdot)\}$ is equicontinuous. Hence, it follows from Theorem 3.3 again that the convergence of $f_n(\cdot)$ to $f(\cdot)$ is uniform.

(ii) $\Rightarrow$ (i): This follows immediately from Theorem 3.3. $\square$

The above result essentially says that under certain conditions, one has

$$\nabla \left[\lim_{n \to \infty} f_n(x) \right] = \lim_{n \to \infty} \nabla f_n(x). \tag{4.11}$$

To conclude this section, we present an interesting result.

Theorem 4.11. (Riemann–Lebesgue Lemma) *Let* $f(\cdot) \in C([a,b]; \mathbb{R})$. *Then the following hold:*

$$\lim_{n \to \infty} \int_a^b f(x) \sin nx dx = 0. \tag{4.12}$$

$$\lim_{n \to \infty} \int_a^b f(x) \cos nx dx = 0. \tag{4.13}$$

Note that the integrands in the above are not convergent as $n \to \infty$. Thus, the above result gives us a quite different flavor. Actually, we have a more general result of which the above is a direct consequence.

Theorem 4.12. (Extended Riemann–Lebesgue Lemma) *Let* $f(\cdot) \in C([a,b]; \mathbb{R})$ *and* $g(\cdot) \in C(\mathbb{R}; \mathbb{R})$ *such that*

$$g(x + T) = g(x), \qquad \forall x \in \mathbb{R}.$$

Then

$$\lim_{n \to \infty} \int_a^b f(x) g(nx) dx = \frac{1}{T} \left(\int_0^T g(x) dx \right) \int_a^b f(x) dx. \tag{4.14}$$

Proof. First, we assume that $f(\cdot)$ is continuously differentiable and

$$\int_0^T g(x) dx = 0.$$

Let

$$G(x) = \int_0^x g(t) dt, \qquad \forall x \in \mathbb{R}.$$

Then $G(\cdot)$ is continuous on $\mathbb{R}$ and periodic with period T. Hence, it is bounded. We obtain

$$\int_a^b f(x) g(nx) dx = \frac{1}{n} \int_a^b f(x) dG(nx)$$

$$= \frac{1}{n} [f(b) G(nb) - f(a) G(na)] - \frac{1}{n} \int_a^b f'(x) G(nx) dx \to 0.$$

Next, we take a sequence of polynomials $p_k(\cdot)$ such that

$$\|f(\cdot) - p_k(\cdot)\|_\infty < \frac{1}{k}.$$

Since $g(\cdot)$ is continuous and periodic, we may let $M = \|g(\cdot)\|_\infty$. Then

$$\left| \int_a^b f(x)g(nx)dx \right| \leqslant \left| \int_a^b p_k(x)g(nx)dx \right| + \int_a^b |g(x)| \, |f(x) - p_k(x)|dx$$

$$\leqslant \left| \int_a^b p_k(x)g(nx)dx \right| + \frac{M(b-a)}{k}.$$

Hence,

$$\varlimsup_{n \to \infty} \left| \int_a^b f(x)g(nx)dx \right| \leqslant \frac{M(b-a)}{k}.$$

Since k could be arbitrarily large, we obtain

$$\lim_{n \to \infty} \left| \int_a^b f(x)g(nx)dx \right| = 0.$$

Finally, for the general case, from the above, we have

$$\lim_{n \to \infty} \int_a^b f(x)\left(g(nx) - \frac{1}{T} \int_0^T g(t)dt \right)dx = 0.$$

Then our result follows. □

Let us look at the following example.

Example 4.13. Let $f(\cdot) \in C([a,b]; \mathbb{R})$. Then

$$\lim_{n \to \infty} \int_a^b f(x)|\sin nx|dx = \frac{2}{\pi} \int_a^b f(x)dx.$$

Since $g(x) = |\sin x|$ has period π, and

$$\frac{1}{\pi} \int_0^\pi |\sin x|dx = \frac{1}{\pi} \int_0^\pi \sin x dx = \frac{2}{\pi}.$$

Similarly,

$$\lim_{n \to \infty} \int_a^b f(x)|\cos nx|dx = \frac{2}{\pi} \int_a^b f(x)dx.$$

Exercises

1. Let $f_n : (a,b) \to \mathbb{R}$ be a sequence of continuous functions such that

$$|f_n(x)| \leqslant M, \qquad \forall n \geqslant 1, \ x \in (a,b),$$

for some constant $M > 0$, and

$$\lim_{n\to\infty} f_n(x) = f(x), \qquad x \in (a, b),$$

for some $f : (a, b) \to \mathbb{R}$. Moreover, for any small $\varepsilon > 0$, the above convergence is uniform on $[a + \varepsilon, b - \varepsilon]$. Then $f(\cdot)$ is continuous on (a, b) and

$$\lim_{n\to\infty} \int_a^b f_n(x)dx = \int_a^b f(x)dx.$$

2. Let $f : [0, \infty) \to [0, \infty)$ be continuous monotone increasing. Find

$$\lim_{n\to\infty} \int_0^1 f\left(\frac{nx}{1 + nx}\right)dx.$$

3. Let $f_n, f, g : \mathbb{R} \to \mathbb{R}$ be Riemann integrable (not necessarily bounded), $n \geqslant 1$. Suppose that

$$|f_n(x)| \leqslant g(x), \qquad \lim_{n\to\infty} f_n(x) = f(x), \qquad x \in \mathbb{R}.$$

Then

$$\lim_{n\to\infty} \int_{\mathbb{R}} |f_n(x) - f(x)|dx = 0.$$

4. State the Levi's and Fatou's Lemmas under weaker conditions and prove them using Arzelà's Theorem.

5. Let $f_n, f : [a, b] \to [0, \infty)$ be continuous such that

$$\inf_{n\geqslant m} \left(f_n - |f_n - f|\right)$$

is continuous for each $m \geqslant 1$ and

$$\lim_{n\to\infty} f_n(x) = f(x), \quad x \in [a, b]; \qquad \lim_{n\to\infty} \int_a^b f_n(x)dx = \int_a^b f(x)dx.$$

Then

$$\lim_{n\to\infty} \int_a^b |f_n(x) - f(x)|dx = 0.$$

Chapter 6

Series

1 Series of Numbers

We first introduce the following definition.

Definition 1.1. (i) Let $\{a_n\} \in \ell^0$. A formal summation $\sum_{n=0}^{\infty} a_n \equiv \sum a_n$ is called a *series*.

(ii) For a series $\sum a_n$, the following is called a *partial sum* of the series:

$$S_N = \sum_{n=0}^{N} a_n, \qquad N \geqslant 0.$$

(iii) A series $\sum a_n$ is said to be *convergent* if the sequence $\{S_N\}_{N\geqslant 0}$ of partial sums is convergent, i.e., there exists an $L \in \mathbb{R}$ such that for any $\varepsilon > 0$, one can find an $N_0 \geqslant 0$ having the following property

$$|S_N - L| \equiv \left| \sum_{n=0}^{N} a_n - L \right| < \varepsilon, \qquad \forall N \geqslant N_0.$$

In this case, L is called the *sum* of the series and denote $L = \sum a_n$. If $\sum a_n$ is not convergent, it is said to be *divergent*.

From the above, we see that a series is convergent if and only if the sequence of its partial sums is convergent. Hence, we have the following result whose proof is left to the readers.

Proposition 1.2. (i) *Series* $\sum a_n$ *is convergent if and only if for any* $\varepsilon > 0$, *there exists an* $N \geqslant 1$ *such that*

$$\left| \sum_{n=p}^{q} a_n \right| < \varepsilon, \qquad \forall q > p \geqslant N.$$

151

(ii) If $\sum a_n$ is convergent, then it is necessary that the *general term* a_n *approaches to 0*, i.e.,

$$\lim_{n \to \infty} a_n = 0.$$

(iii) Series $\sum\limits_{n=0}^{\infty} a_n$ is convergent if and only if for any fixed $m \geqslant 1$, series $\sum\limits_{n=m}^{\infty} a_n$ is convergent.

We see that the convergence of a series $\sum\limits_{n=0}^{\infty} a_n$ depends only on its "tail"; any first finitely many terms are irrelevant to the convergence. If the series is convergent, the first finitely many terms only possibly make the sum of the series different. Therefore, if we are only interested in the convergence of a series, then it suffices to simply write $\sum a_n$, which has been done already in the above, unless the sum of the series is an interested part of the problem.

Example 1.3. (i) *Geometric series* $\sum\limits_{n=0}^{\infty} x^n$ is convergent for $|x| < 1$ and it is divergent for $|x| \geqslant 1$. In fact, for any $x \in \mathbb{R}$,

$$S_N = \sum_{n=0}^{N} x^n = \frac{1 - x^{N+1}}{1 - x}, \qquad N \geqslant 0.$$

If $|x| < 1$, one has

$$\lim_{N \to \infty} S_N = \frac{1}{1 - x}.$$

Hence, $\sum\limits_{n=0}^{\infty} x^n = \dfrac{1}{1-x}$, for $|x| < 1$.

On the other hand, if $|x| \geqslant 1$, then x^n does not go to zero as $n \to \infty$. Hence, the series is divergent. A special case is $x = -1$. Thus, series $\sum (-1)^n$ is divergent.

(ii) *Harmonic series* $\sum\limits_{n=1}^{\infty} \dfrac{1}{n}$ is divergent. For this series, although the general term $\frac{1}{n}$ converges to 0, but the speed is not fast enough. We look at the following:

$$1 + \frac{1}{2} + \frac{1}{3} + \frac{1}{4} + \frac{1}{5} + \frac{1}{6} + \frac{1}{7} + \frac{1}{8} + \frac{1}{9} \cdots$$

$$> 1 + \frac{1}{2} + \left(\frac{1}{4} + \frac{1}{4}\right) + \left(\frac{1}{8} + \frac{1}{8} + \frac{1}{8} + \frac{1}{8}\right) + \cdots$$

$$= 1 + \frac{1}{2} + \frac{1}{2} + \cdots = \infty.$$

In this case,

$$S_1 = 1, \quad S_2 = 1 + \frac{1}{2}, \quad S_4 > 1 + \frac{2}{2}, \quad S_8 > 1 + \frac{3}{2}, \cdots.$$

In general,

$$S_{2^k} > 1 + \frac{k}{2}, \qquad k \geqslant 0.$$

Thus,

$$\lim_{N \to \infty} S_N = \infty.$$

The divergence of the harmonic series tells us that the general term $a_n \to 0$ as $n \to \infty$ is not sufficient for the convergence of the series. Furthermore, we have the following example concerning the Cauchy sequence of real numbers.

Example 1.4. Let $m > 1$ be fixed. There exists a sequence $\{s_n\}_{n \geqslant 1} \in \ell^0$ such that for any $\varepsilon > 0$ and any $m \geqslant 1$, there exists an $N \geqslant 1$ such that

$$|s_k - s_\ell| < \varepsilon, \qquad \forall n \leqslant k, \ell \leqslant n + m, \ n \geqslant N. \tag{1.1}$$

But $\{s_n\}_{n \geqslant 1}$ is not convergent. This means that a sequence satisfies the above condition might not be a Cauchy sequence.

Such a sequence can be constructed in the following way: Let

$$a_{jm+1} = a_{jm+2} = \cdots = a_{(j+1)m} = \frac{1}{(j+1)m}, \qquad j \geqslant 0.$$

Then define

$$s_k = \sum_{i=1}^{k} a_i, \qquad k \geqslant 1.$$

It is clear that

$$s_{jm} = 1 + \frac{1}{2} + \frac{1}{3} + \cdots + \frac{1}{j}, \qquad \forall j \geqslant 1.$$

Thus, $s_k \to \infty$. However, for any $n \geqslant 1$, we may let $jm < n \leqslant (j+1)m$. Then for any $n \leqslant k < \ell \leqslant n + m$, we have

$$0 < s_\ell - s_k < \frac{1}{j}.$$

Thus, (1.1) holds.

A particular case of the above says that if for any $\varepsilon > 0$, there exists an $N \geqslant 1$ such that

$$|s_{n+1} - s_n| < \varepsilon, \qquad \forall n \geqslant N,$$

we do not necessarily have the Cauchy property of such a sequence $\{s_n\}$.

Definition 1.5. (i) Series $\sum a_n$ is said to be *absolutely convergent* if $\sum |a_n|$ is convergent.

(ii) Series $\sum a_n$ is said to be *conditionally convergent* if it is convergent but not absolutely convergent.

Proposition 1.6. (i) *If $\sum a_n$ is absolutely convergent, then it is convergent.*

(ii) *If there exists a $K \geqslant 0$ such that*

$$a_n \geqslant 0, \qquad \forall n \geqslant K,$$

then $\displaystyle\sum_{n=0}^{\infty} a_n$ is convergent if and only if there exists an $M > 0$ such that

$$\sum_{n=0}^{N} a_n \leqslant M, \qquad \forall N \geqslant 1.$$

Further, in this case, the series must also be absolutely convergent.

(iii) **(Comparison Test)** *Suppose there exists an $N \in \mathbb{N}$ such that*

$$|a_n| \leqslant b_n, \qquad n \geqslant N,$$

and $\sum b_n$ is convergent. Then $\sum a_n$ is absolutely convergent.

(iv) **(Ratio Comparison Test)** *Let $\sum a_n$ and $\sum b_n$ be two series of positive numbers. Suppose that there exists an $N \geqslant 1$ such that*

$$\frac{a_{n+1}}{a_n} \leqslant \frac{b_{n+1}}{b_n}, \qquad \forall n \geqslant N. \tag{1.2}$$

Then the convergence of $\sum b_n$ implies that of $\sum a_n$.

(v) **(Alternative Terms Test)** *Suppose $\{a_n\}$ is a decreasing sequence. Then $\sum (-1)^n a_n$ is convergent if and only if $a_n \to 0$.*

(vi) *There exists a conditionally convergent series.*

Proof. (i) For any $\varepsilon > 0$, there exists an $N \geqslant 1$ such that

$$\left| \sum_{n=p}^{q} a_n \right| \leqslant \sum_{n=p}^{q} |a_n| < \varepsilon, \qquad \forall q > p \geqslant N.$$

This proves our claim.

(ii) For any $N > K$,

$$S_N = S_K + \sum_{n=K+1}^{N} a_n.$$

It is clear that $\{S_N\}_{N \geqslant K+1}$ is a non-decreasing sequence. Hence, it is convergent if and only if S_N is bounded from above.

(iii) For any $\varepsilon > 0$, there exists an $N \geqslant 1$ such that

$$\sum_{n=p}^{q} b_n < \varepsilon, \qquad \forall q > p \geqslant N.$$

Then

$$\sum_{n=p}^{q} |a_n| \leqslant \sum_{n=p}^{q} b_n < \varepsilon, \qquad q > p \geqslant N.$$

This leads to the absolute convergence of $\sum a_n$.

(iv) Without loss of generality, we let $N = 1$. Then

$$a_n = a_1 \cdot \frac{a_2}{a_1} \cdot \frac{a_3}{a_2} \cdots \frac{a_n}{a_{n-1}} \leqslant a_1 \frac{b_2}{b_1} \cdot \frac{b_3}{b_2} \cdots \frac{b_n}{b_{n-1}} = \frac{a_1}{b_1} b_n.$$

Hence, by the convergence of $\sum b_n$ we see that $\sum a_n$ is convergent.

(v) Note that

$$S_{2k+2} = S_{2k} + (-1)^{2k+1} a_{2k+1} + (-1)^{2k+2} a_{2k+2}$$
$$= S_{2k} - (a_{2k+1} - a_{2k+2}) \leqslant S_{2k}, \qquad \forall k \geqslant 0$$

and

$$S_{2k+1} = S_{2k-1} + (-1)^{2k} a_{2k} + (-1)^{2k+1} a_{2k+1}$$
$$= S_{2k-1} + a_{2k} - a_{2k+1} \geqslant S_{2k-1}, \qquad \forall k \geqslant 0.$$

Also,

$$S_{2k+2} = S_{2k+1} + a_{2k+2} \geqslant S_{2k+1} \geqslant S_{2k-1} \geqslant \cdots \geqslant S_1,$$

and

$$S_{2k+1} = S_{2k} - a_{2k+1} \leqslant S_{2k} \leqslant S_{2k-2} \leqslant \cdots \leqslant S_0.$$

Hence, $\{S_{2k}\}_{k \geqslant 0}$ is decreasing, bounded below, and $\{S_{2k+1}\}_{k \geqslant 0}$ is increasing, bounded above. Therefore,

$$\lim_{k \to \infty} S_{2k} = S, \qquad \lim_{k \to \infty} S_{2k+1} = \bar{S}.$$

Further,

$$|S_{2k+1} - S_{2k}| = a_{2k+1}, \qquad \forall k \geqslant 0,$$

which implies that $S = \bar{S}$ if and only if $a_n \to 0$.

(vi) Let $a_n = \frac{(-1)^n}{n}$. Then $\displaystyle\sum_{n=1}^{\infty} a_n$ is conditionally convergent. $\qquad\square$

The following collects some basic properties of convergent series.

Proposition 1.7. (i) *Suppose* $\sum a_n$ *and* $\sum b_n$ *are convergent. Then*

$$\sum (a_n + b_n) = \sum a_n + \sum b_n, \qquad \sum (ca_n) = c \sum a_n, \qquad \forall c \in \mathbb{R}.$$

(ii) $\displaystyle\sum_{n=0}^{\infty} a_n$ *is convergent if and only if* $\displaystyle\sum_{n=k}^{\infty} a_{n-k}$ *is convergent. In this case,*

$$\sum_{n=0}^{\infty} a_n = \sum_{n=k}^{\infty} a_{n-k}.$$

(iii) *Let* $a_n \to 0$. *Then*

$$\sum_{n=0}^{\infty} (a_n - a_{n+1}) = a_0.$$

The proof is left to the readers.

Exercises

1. Let $\sum a_n$ be convergent. Must $\sum (-1)^n a_n$ also be convergent? Why?

2. Let $\sum a_n$ be convergent with all $a_n > 0$. Is it possible that $\sum \frac{1}{a_n}$ is convergent? Why? How about $\sum \frac{a_n}{1+a_n}$? Why?

3. Let $\sum a_n$ and $\sum b_n$ be convergent. Is $\sum a_n b_n$ convergent? Why? How about $\sum \frac{a_n}{b_n}$ if all $b_n \neq 0$? Why?

4. Let $\displaystyle\sum_{n=1}^{\infty} a_n$ be convergent. Let $b_n = \dfrac{1}{n}\displaystyle\sum_{k=1}^{n} a_k$. Is $\sum b_n$ convergent? Why?

5. Let $a_n \geqslant 0$, and $a_n \to 0$. Let $m > 1$ be fixed. Define

$$b_{mk+j} = (-1)^k a_{mk+j}, \qquad k \geqslant 0, \quad 0 \leqslant j \leqslant m - 1.$$

Is $\sum b_n$ convergent? Why?

2 Further Tests for Convergence

We now present some further tests for the convergence of series.

Proposition 2.1. (Cauchy Test) *Let*

$$a_n \geqslant a_{n+1} > 0, \qquad \forall n \geqslant 1.$$

Then $\displaystyle\sum_{n=1}^{\infty} a_n$ *is convergent if and only if* $\displaystyle\sum_{k=0}^{\infty} 2^k a_{2^k}$ *is convergent.*

Proof. Let

$$S_N = \sum_{n=1}^{N} a_n, \qquad T_K = \sum_{k=0}^{K} 2^k a_{2^k}, \qquad \forall N \geqslant 1, \ K \geqslant 0.$$

We now prove the following by induction:

$$S_{2^{K+1}-1} \leqslant T_K \leqslant 2S_{2^K}, \qquad \forall K \geqslant 0. \tag{2.1}$$

For $K = 0$, we need

$$S_1 \leqslant T_0 \leqslant 2S_1,$$

which is the following true statement:

$$a_1 \leqslant a_1 \leqslant 2a_1.$$

Suppose our claim (2.1) is true for some K. Then, for $K + 1$, we need to prove

$$S_{2^{K+2}-1} \leqslant T_{K+1} \leqslant 2S_{2^{K+1}}. \tag{2.2}$$

Note that

$$T_{K+1} = T_K + 2^{K+1} a_{2^{K+1}}. \tag{2.3}$$

Also,

$$S_{2^{K+1}} = S_{2^K} + \sum_{n=2^K+1}^{2^{K+1}} a_n \geqslant S_{2^K} + \sum_{n=2^K+1}^{2^{K+1}} a_{2^{K+1}} = S_{2^K} + 2^K a_{2^{K+1}}.$$

Hence, by induction hypothesis, one has

$$2S_{2^{K+1}} \geqslant 2S_{2^K} + 2^{K+1}a_{2^{K+1}} \geqslant T_K + 2^{K+1}a_{2^{K+1}} = T_{K+1}. \qquad (2.4)$$

Similarly, (noting (2.3), and induction hypothesis)

$$
\begin{aligned}
S_{2^{K+2}-1} &= S_{2^{K+1}-1} + \sum_{n=2^{K+1}}^{2^{K+2}-1} a_n \leqslant S_{2^{K+1}-1} + \sum_{n=2^K+1}^{2^{K+2}-1} a_{2^{K+1}} \\
&= S_{2^{K+1}-1} + 2^{K+1}a_{2^{K+1}} \leqslant T_K + 2^{K+1}a_{2^{K+1}} = T_{K+1}.
\end{aligned}
\qquad (2.5)
$$

Then, combining (2.4) and (2.5), we obtain (2.2). This proves (2.1). Hence, T_K is bounded from above if and only if S_N is bounded from above. This leads to our proposition. $\qquad \square$

The above result has a very interesting corollary.

Corollary 2.2. *Let $q > 0$ be a real number. Then $\displaystyle\sum_{n=1}^{\infty} \frac{1}{n^q}$ is convergent if and only if $q > 1$.*

Proof. Let

$$a_n = \frac{1}{n^q}, \qquad n \geqslant 1.$$

Then,

$$a_{2^k} = \frac{1}{(2^k)^q}, \qquad \forall k \geqslant 0.$$

Hence, by Proposition 2.1, we see that $\displaystyle\sum_{n=1}^{\infty} \frac{1}{n^q}$ is convergent if and only if

$$\sum_{k=0}^{\infty} 2^k \frac{1}{(2^k)^q} = \sum_{k=0}^{\infty} \frac{1}{(2^{q-1})^k}$$

is convergent. For such a geometric series, it is convergent if and only if $q > 1$. $\qquad \square$

The following is another relevant test.

Theorem 2.3. (Cauchy's Integral Test) *Let $f : (0, \infty) \to (0, \infty)$ be continuous and non-increasing on $[N, \infty)$ for some $N \in \mathbb{N}$. Let*

$$a_n = f(n), \qquad n \geqslant 1.$$

Then $\sum a_n$ is convergent if and only if

$$\lim_{b \to \infty} \int_N^b f(x)dx < \infty. \qquad (2.6)$$

Proof. For any $n \geqslant N$, we have

$$a_{n+1} = f(n+1) \leqslant f(x) \leqslant f(n) = a_n, \qquad \forall x \in [n, n+1].$$

Thus,

$$a_{N+1} + \cdots + a_{n+1} \leqslant \int_N^{n+1} f(x)dx \leqslant a_N + \cdots + a_n.$$

Consequently, our conclusion follows. $\qquad\qquad\qquad\qquad\square$

The above Corollary 2.2 can also be regarded as a consequence of the above Cauchy's integral test.

Let us briefly mention an interesting problem relevant to the series in Corollary 2.2.[1] To this end, we recall (from the Appendix) $\mathbb{C}$, the set of all complex numbers. We look at the following series:

$$\zeta(s) = \sum_{n=1}^{\infty} \frac{1}{n^s}, \tag{2.7}$$

where $s \in \mathbb{C}$. Write $s = \sigma + it$ with $\sigma, t \in \mathbb{R}$. Then

$$|n^{-s}| = |n^{-(\sigma+it)}| = |n^{-\sigma}||e^{it \log n}| = \frac{1}{n^{\sigma}}.$$

Thus, series (2.7) is absolutely convergent for $\sigma = \operatorname{Re} s > 1$. One can show that the above function admits an *analytic extension* on the complex plan $\mathbb{C}$, still denote it by $\zeta(s)$, except for the point $s = 1$, with

$$\lim_{s \to 1}(s-1)\zeta(s) = 1.$$

Such an extension is called the *Riemann zeta function*. One can show that function $\zeta(\cdot)$ satisfies the following *Riemann's functional equation*:

$$\zeta(s) = 2^s \pi^{s-1} \sin\left(\frac{\pi s}{2}\right)\Gamma(1-s)\zeta(1-s), \qquad s \in \mathbb{C} \setminus \{1\}, \tag{2.8}$$

where $\Gamma(\cdot)$ is called the *Gamma function* defined by the following:

$$\Gamma(s) = \int_0^{\infty} e^{-x} x^{s-1} dx.$$

Note that in the above functional equation,

$$\lim_{s \to 2n} \sin\left(\frac{\pi s}{2}\right)\Gamma(1-s) \neq 0, \qquad n \geqslant 0,$$

and

$$\zeta(-2n) = 0, \qquad n \geqslant 1.$$

---

[1] Since this part of the material is not closely related to the following material, the first time reader of the book could skip this part.

Thus, all negative even numbers are called the *trivial zeros* of $\zeta(\cdot)$. The well-known *Riemann hypothesis* says that:

All the non-trivial zeros of $\zeta(\cdot)$ lie on the line $\operatorname{Re} s = \frac{1}{2}$.

This is a very important and challenging conjecture. It is open as of today.

Example 2.4. Consider series $\sum_{n=0}^{\infty} \frac{1}{n!}$. Let

$$S_n = \frac{1}{0!} + \frac{1}{1!} + \frac{1}{2!} + \cdots + \frac{1}{n!} = \sum_{k=0}^{n} \frac{1}{k!}, \qquad n \geqslant 0, \qquad (2.9)$$

since each term in the sum is positive, we see that S_n is increasing. We now show that S_n is bounded from above. In fact, for any $n > 2$, we have

$$\frac{1}{n!} < \frac{1}{2^{n-1}}. \qquad (2.10)$$

Thus,

$$S_n < 1 + 1 + \frac{1}{2} + \cdots + \frac{1}{2^{n-1}} = 1 + \frac{1 - (\frac{1}{2})^n}{1 - \frac{1}{2}} < 3, \qquad \forall n \geqslant 1. \qquad (2.11)$$

Hence, sequence $\{S_n\}_{n \geqslant 1}$ is convergent.

Proposition 2.5. (Cauchy's Root Test) *For series* $\sum a_n$, *let*

$$\alpha = \varlimsup_{n \to \infty} |a_n|^{\frac{1}{n}}.$$

Then

(i) $\alpha < 1$, *series is absolutely convergent;*

(ii) $\alpha > 1$, *series is divergent;*

(iii) $\alpha = 1$, *the test fails.*

Proof. In the case that $0 < \alpha < 1$, we can find an $N \geqslant 1$ such that

$$|a_n|^{\frac{1}{n}} \leqslant \sup_{m \geqslant n} |a_m|^{\frac{1}{m}} < \alpha + \frac{1 - \alpha}{2} = \frac{\alpha + 1}{2} < 1, \qquad \forall n \geqslant N.$$

Hence,

$$\sum |a_n| = \sum_{n \leqslant N} |a_n| + \sum_{n=N+1}^{\infty} (|a_n|^{\frac{1}{n}})^n \leqslant \sum_{n \leqslant N} |a_n| + \sum_{n=N+1}^{\infty} \left(\frac{\alpha + 1}{2}\right)^n < \infty.$$

In the case $\alpha > 1$, we have a subsequence a_{n_k} such that

$$|a_{n_k}|^{\frac{1}{n_k}} > \alpha - \frac{\alpha - 1}{2} = \frac{\alpha + 1}{2}, \qquad k \geqslant 1.$$

Hence,

$$|a_{n_k}| \geqslant \left(\frac{\alpha+1}{2}\right)^{n_k} \geqslant 1, \qquad \forall k \geqslant 1,$$

which implies that $|a_n|$ does not go to 0. Consequently, $\sum a_n$ is divergent.

Finally, in the case $\alpha = 1$, the series could be convergent and also could be divergent. For example,

$$\sum_{n=1}^{\infty} \frac{1}{n^2} < \infty, \qquad \lim_{n\to\infty} \left(\frac{1}{n^2}\right)^{\frac{1}{n}} = 1,$$

and

$$\sum_{n=1}^{\infty} \frac{1}{n} = \infty, \qquad \lim_{n\to\infty} \left(\frac{1}{n}\right)^{\frac{1}{n}} = 1.$$

This completes the proof. □

Theorem 2.6. (d'Alembert's Ratio Test) *For series $\sum a_n$, let $a_n \neq 0$ for all $n \geqslant N$ with some $N \in \mathbb{N}$ and let*

$$\alpha = \overline{\lim_{n\to\infty}} \frac{|a_{n+1}|}{|a_n|}, \qquad \beta = \underline{\lim_{n\to\infty}} \frac{|a_{n+1}|}{|a_n|}.$$

Then

    (i) $\alpha < 1$, *the series is absolutely convergent;*

    (ii) $\beta > 1$, *the series is divergent;*

    (iii) $\alpha = 1$, *or $\beta = 1$, the test fails.*

Proof. In the case that $\alpha < 1$, we have $N \geqslant m$ such that

$$\frac{|a_{n+1}|}{|a_n|} \leqslant \sup_{m\geqslant n} \frac{|a_{m+1}|}{|a_m|} < \alpha + \frac{1-\alpha}{2} = \frac{1+\alpha}{2} < 1, \qquad \forall n \geqslant N.$$

Hence,

$$|a_n| \leqslant \left(\frac{1+\alpha}{2}\right)^{n-N} |a_N|, \qquad \forall n \geqslant N.$$

This leads to

$$\sum |a_n| = \sum_{n\leqslant N} |a_n| + \sum_{n=N+1}^{\infty} |a_n|$$

$$\leqslant \sum_{n\leqslant N} |a_n| + \sum_{n=N+1}^{\infty} \left(\frac{1+\alpha}{2}\right)^{n-N} |a_N| < \infty.$$

Next, for the case $\beta > 1$, we have $N \geqslant m$ such that

$$\frac{|a_{n+1}|}{|a_n|} \geqslant \inf_{m \geqslant n} \frac{|a_{m+1}|}{|a_m|} > \beta - \frac{\beta - 1}{2} = \frac{1 + \beta}{2} > 1, \qquad \forall n \geqslant N.$$

Hence,

$$|a_n| \geqslant \left(\frac{1 + \beta}{2}\right)^{n-N} |a_N|, \qquad n \geqslant N,$$

and

$$\lim_{n \to \infty} |a_n| = \infty.$$

Consequently, the series is divergent.

Finally,

$$\sum_{n=1}^{\infty} \frac{1}{n^2} < \infty, \qquad \lim_{n \to \infty} \frac{1/(n+1)^2}{1/n^2} = 1,$$

and

$$\sum_{n=1}^{\infty} \frac{1}{n} = \infty, \qquad \lim_{n \to \infty} \frac{1/(n+1)}{1/n} = 1.$$

Thus, in the case $\alpha = 1$ or $\beta = 1$, the test fails. $\qquad\square$

We have seen that when

$$\lim_{n \to \infty} \left|\frac{a_{n+1}}{a_n}\right| = 1, \tag{2.12}$$

the d'Alembert's ratio test fails. The following gives some further test when (2.12) holds.

Theorem 2.7. (Raabe's Test)[*] *Suppose that*

$$t = \lim_{n \to \infty} \left\{ n\left(1 - \left|\frac{a_{n+1}}{a_n}\right|\right) \right\} \tag{2.13}$$

exists (including $\pm\infty$). Then $\sum a_n$ is absolutely convergent if $t > 1$, and not absolutely convergent if $t < 1$.

Proof. First of all, applying the Taylor's formula to $f(x) = (1 + x)^{-p}$, for some $p > 0$, at $x = 0$, we get

$$(1 + x)^{-p} = 1 - px + \frac{p(p + 1)x^2}{2(1 + \theta x)^{p+2}}, \qquad \theta \in (0, 1).$$

Hence, letting $x = \frac{1}{n}$, one has

$$\left(1 + \frac{1}{n}\right)^{-p} = 1 - \frac{p}{n} + \frac{A_n}{n^2}, \qquad A_n \equiv \frac{p(p + 1)}{2(1 + \frac{\theta}{n})^{p+2}} > 0. \tag{2.14}$$

Now, suppose $t > 1$ (allowing $t = \infty$). Let $p \in (1, t)$. Then there exists an $N \geqslant 1$ such that

$$p < n\left(1 - \left|\frac{a_{n+1}}{a_n}\right|\right), \qquad n \geqslant N,$$

which, together with (2.14), implies that

$$\left|\frac{a_{n+1}}{a_n}\right| < 1 - \frac{p}{n} < \left(1 + \frac{1}{n}\right)^{-p} = \frac{(n+1)^{-p}}{n^{-p}}, \qquad \forall n \geqslant N.$$

Hence, by Proposition 1.6 (iv), and the convergence of $\sum n^{-p}$ (note $p > 1$), we obtain the convergence of $\sum a_n$.

Now, let $t < 1$. We can take $p \in (t, 1)$. Since $A_n \geqslant 0$ is bounded, we may find some $N \geqslant 1$ large enough so that

$$p - \frac{A_n}{n} > n\left(1 - \left|\frac{a_{n+1}}{a_n}\right|\right), \qquad n \geqslant N,$$

which, together with (2.14), implies

$$\left|\frac{a_{n+1}}{a_n}\right| > 1 - \frac{p}{n} + \frac{A_n}{n^2} = \left(1 + \frac{1}{n}\right)^{-p} = \frac{(n+1)^{-p}}{n^{-p}}.$$

Hence, by Proposition 1.6 (iv) again, and the convergence of $\sum |a_n|$ will lead to the convergence of $\sum n^{-p}$ which is not the case since $p < 1$. Therefore, $\sum |a_n|$ must be divergent. $\qquad\square$

If in the above, $t = 1$, then the test fails. In this case, one has

$$\left|\frac{a_{n+1}}{a_n}\right| \sim 1 - \frac{1}{n}.$$

The following says more for such a case.

Theorem 2.8. (Gauss' Test)* *Let*

$$\left|\frac{a_{n+1}}{a_n}\right| = 1 - \frac{p}{n} + \frac{A_n}{n^q}, \qquad (2.15)$$

where $q > 1$ and $\{A_n\}$ is bounded. Then $\sum a_n$ is absolutely convergent if $p > 1$, and not absolutely convergent (either divergent or conditionally convergent) if $p \leqslant 1$.

Proof. By (2.15), we see that

$$\lim_{n \to \infty} \left\{ n\left(1 - \left|\frac{a_{n+1}}{a_n}\right|\right) \right\} = p. \qquad (2.16)$$

Thus, the conclusions for $p \neq 1$ follows from the Raabe's test.

We now look at the case $p = 1$. In order to consider this case, we first apply the Taylor's formula to $f(x) = \log(1 - x)$ at $x = 0$ to get

$$\log(1 - x) = -x - \frac{x^2}{2(1 - \theta x)^2}, \qquad \theta \in (0, 1).$$

Taking $x = \frac{1}{n}$, we have

$$\log(1 - \frac{1}{n}) = -\frac{1}{n} - \frac{\widetilde{A}_n}{n^2}, \qquad \widetilde{A}_n = \frac{1}{2(1 - \frac{\theta}{n})^2}. \tag{2.17}$$

Next, let

$$b_n = \frac{1}{(n - 1)\log(n - 1)}, \qquad n \geqslant 2.$$

Then

$$\frac{b_{n+1}}{b_n} = \frac{(n - 1)\log(n - 1)}{n \log n} = \frac{(n - 1)[\log n + \log(1 - \frac{1}{n})]}{n \log n}$$

$$= 1 - \frac{1}{n} - \frac{(n - 1)}{n \log n}\left(\frac{1}{n} + \frac{\widetilde{A}_n}{n^2}\right)$$

$$= 1 - \frac{1}{n} - \frac{1}{n \log n} + \frac{B_n}{n^2},$$

with

$$B_n \equiv \frac{1}{\log n} - \frac{\widetilde{A}_n(n - 1)}{n \log n},$$

which is bounded. Then we have

$$\left|\frac{a_{n+1}}{a_n}\right| - \frac{b_{n+1}}{b_n} = \frac{1}{n \log n} + \frac{A_n}{n^q} - \frac{B_n}{n^2} \geqslant 0,$$

when n is large enough. Since $\sum b_n$ is divergent, we have the divergence of $\sum |a_n|$.   □

For any finite sequence $\{a_n\}_{n=1}^m$ of real numbers, let $\tau : \{1, 2, \cdots, m\} \to \{1, 2, \cdots, m\}$ be any bijection. Then

$$\sum_{n=1}^m a_n = \sum_{n=1}^m a_{\tau(n)}.$$

We call $\{a_{\tau(n)}\}_{n=1}^m$ a *rearrangement* of $\{a_n\}_{n=1}^m$. The above means that the sum of a rearrangement of a finite series is the same as that of the original (finite) series. The question is whether this is true for infinite series.

Example 2.9. Consider series $\displaystyle\sum_{n=1}^{\infty} \frac{(-1)^n}{n}$ which is conditionally convergent. Note that

$$\frac{1}{2} + \frac{1}{4} + \cdots + \frac{1}{2N} = \sum_{k=1}^N \frac{1}{2k} = \frac{1}{2}\sum_{k=1}^N \frac{1}{k} \to \infty, \qquad N \to \infty.$$

Hence, there exists an $N_i \geqslant 1$ such that

$$\sum_{k=1}^{N_1} \frac{1}{2k} > 1, \quad \sum_{k=N_1+1}^{N_2} \frac{1}{2k} > 2, \quad \cdots .$$

In general,

$$\sum_{k=N_i+1}^{N_{i+1}} \frac{1}{2k} > i, \quad i \geqslant 1.$$

Then we make the following rearrangement:

$$\frac{1}{2}, \cdots, \frac{1}{2N_1}, \quad -1,$$
$$\frac{1}{2(N_1+1)}, \cdots, \frac{1}{2N_2}, \quad -\frac{1}{3},$$
$$\cdots$$
$$\frac{1}{2(N_i+1)}, \cdots, \frac{1}{2N_{i+1}}, \quad -\frac{1}{(2i+1)}, \cdots .$$

The resulting series is divergent.

Actually, we have the following result.

Proposition 2.10. *Let* $\displaystyle\sum_{n=1}^{\infty} a_n$ *be absolutely convergent, and* $\tau : \mathbb{N} \to \mathbb{N}$ *be any bijection. Then* $\displaystyle\sum_{n=1}^{\infty} a_{\tau(n)}$ *is also absolutely convergent, and*

$$\sum_{n=1}^{\infty} a_n = \sum_{n=1}^{\infty} a_{\tau(n)}.$$

Proof. Let

$$\sum_{n=1}^{\infty} |a_n| = L.$$

For any $n \geqslant 0$, let

$$\bar{\tau}(n) = \max\{\tau(k) \mid 1 \leqslant k \leqslant n\}.$$

Then

$$\sum_{k=1}^{n} |a_{\tau(k)}| \leqslant \sum_{k=1}^{\bar{\tau}(n)} |a_k| \leqslant \sum_{n=1}^{\infty} |a_n| = L.$$

Hence, $\displaystyle\sum_{n=1}^{\infty} |a_{\tau(n)}|$ is convergent, and

$$L' = \sum_{n=1}^{\infty} |a_{\tau(n)}| \leqslant L.$$

By switching the positions of $\displaystyle\sum_{n=1}^{\infty} |a_n|$ and $\displaystyle\sum_{n=1}^{\infty} |a_{\tau(n)}|$, we see that

$$L \leqslant L'.$$

Consequently, our conclusion follows. $\square$

By the similar idea of Example 2.9, we have the following result.

Proposition 2.11. *Let* $\displaystyle\sum_{n=1}^{\infty} a_n$ *be conditionally convergent. Then for any* $L \in \mathbb{R}$, *there exists a bijection* $\tau : \mathbb{N} \to \mathbb{N}$ *such that* $\displaystyle\sum_{n=1}^{\infty} a_{\tau(n)}$ *is convergent and*

$$\sum_{n=1}^{\infty} a_{\tau(n)} = L.$$

We leave the detailed proof to the interested readers.

Exercises

1. Let $\sum a_n$ and $\sum b_n$ be two series of positive terms. Suppose that there exists an $N \geqslant 1$ such that

$$\frac{a_{n+1}}{a_n} \leqslant \frac{b_{n+1}}{b_n}, \qquad \forall n \geqslant N.$$

Then the divergence of $\sum a_n$ implies the divergence of $\sum b_n$.

2. Let $a_n = \dfrac{1}{n^p (\ln n)^q}$. Find necessary and sufficient conditions on p, q such that the series $\sum a_n$ is convergent.

3. Show that for any $a \in \mathbb{R}$, $\displaystyle\sum_{n=1}^{\infty} \frac{a^{2n+1}}{n!}$ is convergent.

4. Suppose $\sum a_n^2$ and $\sum b_n^2$ are convergent. Then $\sum a_n b_n$ is absolutely convergent.

5. Is $\sum \frac{(-1)^n n}{n^2 + 1}$ convergent? absolutely convergent? Why?

6. Let $\sum a_n^2$ be convergent and $p > \frac{1}{2}$. Then $\sum \frac{a_n}{n^p}$ is absolutely convergent.

7. Suppose $\sum a_n$ is convergent. Is $\sum a_n^2$ necessarily convergent? Why?

3   Series of Functions

Let (X, d) be a metric space and $f_n : X \to \mathbb{R}$ be a sequence of functions $n \geqslant 1$. We now consider the following type series: $\sum_{n=1}^{\infty} f_n(x)$. For any given $x \in X$, the above is a usual series of real numbers. To study the above series, we define the *partial sum* as follows:

$$S_N(x) = \sum_{n=1}^{N} f_n(x), \qquad x \in X.$$

We have the following definition.

Definition 3.1. If for each $x \in X$, $S_N(x)$ is convergent to some number, denoted by $S(x)$, then we say that the series is *pointwise convergent*, and write

$$S(x) = \lim_{N \to \infty} S_N(x) = \sum_{n=1}^{\infty} f_n(x), \qquad x \in X.$$

If the convergence is uniform in $x \in X$, we say that the series is *uniformly convergent*.

Definition 3.2. For any bounded function $f : X \to \mathbb{R}$, we define the *sup norm* of $f(\cdot)$ by the following:

$$\|f\|_\infty = \sup_{x \in X} |f(x)|.$$

The following is the simplest criterion for the convergence of function series.

Theorem 3.3. (Weierstrass M-Test) *Let (X, d) be a metric space and let $f_n(\cdot)$ be a sequence of bounded real-valued continuous functions on X such that*

$$\sum_{n=1}^{\infty} \|f_n\|_\infty < \infty. \tag{3.1}$$

Then the series $\sum_{n=1}^{\infty} f_n(x)$ converges uniformly to some continuous function $S(\cdot)$ on X.

Proof. Since

$$\sum_{n=1}^{N} |f_n(x)| \leqslant \sum_{n=1}^{\infty} \|f_n\|_\infty \equiv M < \infty, \quad x \in X,$$

we see that $\sum f_n(x)$ is absolutely converges for each $x \in X$. Next, let $S_N(\cdot)$ be the partial sum sequence. Then $S_N(\cdot)$ is continuous, and for any $N \geqslant 1$ and $m \geqslant 1$, we have

$$\|S_{N+m} - S_N\|_\infty \leqslant \sum_{n=N+1}^{N+m} \|f_n\|_\infty \leqslant \sum_{n=N+1}^{\infty} \|f_n\|_\infty \to 0, \quad N \to \infty.$$

This means that $S_N(\cdot)$ is uniformly Cauchy. Hence, it converges to $S(\cdot)$ uniformly with $S(\cdot)$ being continuous. □

The following is an interesting application of the above Weierstrass M-test, which gives an example of a continuous function that is nowhere differentiable.

Example 3.4. Consider the following series:

$$f(x) = \sum_{n=1}^{\infty} \frac{\cos(8^n \pi x)}{2^n}, \quad x \in \mathbb{R}.$$

Since

$$\left| \frac{\cos(8^n \pi x)}{2^n} \right| \leqslant \frac{1}{2^n}, \quad \forall n \geqslant 1, \ x \in \mathbb{R},$$

by Weierstrass M-test, we see that the series is uniformly convergent. Thus, $f(\cdot)$ is continuous. Now, we want to show that $f(\cdot)$ is nowhere differentiable. To this end, for any $m, j \geqslant 1$, we evaluate

$$\left| f\left(\frac{j+1}{8^m}\right) - f\left(\frac{j}{8^m}\right) \right| = \left| \sum_{n=1}^{\infty} 2^{-n} \left[\cos\left(8^n \pi \frac{j+1}{8^m}\right) - \cos\left(8^n \pi \frac{j}{8^m}\right) \right] \right|$$

$$= \left| \sum_{n=1}^{m-1} 2^{-n} \left[\cos\left(\frac{(j+1)\pi}{8^{m-n}}\right) - \cos\left(\frac{j\pi}{8^{m-n}}\right) \right] \right.$$

$$+ 2^{-m} \left[\cos\left((j+1)\pi\right) - \cos\left(j\pi\right) \right]$$

$$\left. + \sum_{n=m+1}^{\infty} 2^{-n} \left[\cos\left(8^{n-m}(j+1)\pi\right) - \cos\left(8^{n-m}j\pi\right) \right] \right|$$

$$\geqslant 2 \cdot 2^{-m} - \sum_{n=1}^{m-1} \frac{2^{-n}\pi}{8^{m-n}} = 2 \cdot 2^{-m} - \frac{\pi}{8^m} \sum_{n=1}^{m-1} 4^n$$

$$= 2 \cdot 2^{-m} - \frac{\pi}{8^m} \frac{4^m - 4}{4-1} \geqslant \left(2 - \frac{\pi}{3}\right) \cdot 2^{-m} > 0.$$

Now, for any $x_0 \in \mathbb{R}$, we can find $m, j \geq 1$ such that

$$j \leq 8^m x_0 \leq j + 1,$$

which is equivalent to

$$x_0 \in \left[\frac{j}{8^m}, \frac{j+1}{8^m}\right].$$

If $f(\cdot)$ is differentiable at x_0, then for any $\varepsilon > 0$, there exists an $m_0 \geq 1$ such that for $m \geq m_0$ with a proper j,

$$\begin{aligned}
\left(2 - \frac{\pi}{3}\right) 2^{-m} &\leq \left| f\left(\frac{j+1}{8^m}\right) - f\left(\frac{j}{8^m}\right) \right| \\
&\leq \left| f\left(\frac{j+1}{8^m}\right) - f(x_0) \right| + \left| f(x_0) - f\left(\frac{j}{8^m}\right) \right| \\
&\leq (|f'(x_0)| + \varepsilon)\left(\frac{j+1}{8^m} - x_0\right) + (|f'(x_0)| + \varepsilon)\left(x_0 - \frac{j}{8^m}\right) \\
&\leq (|f'(x_0)| + \varepsilon) 8^{-m}.
\end{aligned}$$

This leads to a contradiction.

The following results are concerned with the so-called *integration term-by-term* and *differentiation term-by-term* for the series.

Theorem 3.5. (i) *Let $G \subseteq \mathbb{R}^m$ be a bounded domain with $|\partial G| = 0$. Let $f_n : \bar{G} \to \mathbb{R}$ be a sequence of continuous functions and $\sum_{n=1}^{\infty} f_n(x)$ be uniformly convergent. Then*

$$\int_G \sum_{n=1}^{\infty} f_n(x) dx = \sum_{n=1}^{\infty} \int_G f_n(x) dx.$$

(ii) *Let $f_n : G \to \mathbb{R}$ be a sequence of continuously differentiable functions. Let $\sum_{n=1}^{\infty} \nabla f_n(x)$ be uniformly convergent, and there exists an $x_0 \in G$ such that $\sum_{n=1}^{\infty} f_n(x_0)$ is convergent. Then*

$$\nabla \left(\sum_{n=1}^{\infty} f_n(x) \right) = \sum_{n=1}^{\infty} \nabla f_n(x).$$

Proof. (i) Let $S_N(\cdot)$ be the sequence of partial sums. Now, in the case that $S_N(\cdot)$ converges to $S(\cdot)$ uniformly, we have

$$\int_G S(x) dx = \lim_{N \to \infty} \int_G S_N(x) dx,$$

which is what we want to prove.

(ii) Since $\nabla f(x) = (f_{x^1}(x), \cdots, f_{x^m}(x))$, it suffices to prove the case $m = 1$. Thus, we let $G = (a, b)$. By assumption, $S'_N(\cdot)$ converges to some $g(\cdot)$ uniformly and $S_N(x_0)$ is convergent. Hence, we have

$$S(x) = S(x_0) + \int_{x_0}^{x} g(t)dt.$$

Consequently,

$$S'(x) = g(x), \qquad x \in [a, b],$$

which is exactly what we want.   □

Note that if we use Arzelà Theorem (Theorem 4.5 of Chapter 5), the above can be improved.

To conclude this section, we look at limit theorems for sequences of number series.

Theorem 3.6. *Let $\{a_{nm}\}_{n,m\geqslant 1}$ be a double-indexed sequence of real numbers.*

(i) **(Dominated Convergence Theorem)** *Let there exist sequences $\{b_n\}_{n\geqslant 1}$ and $\{c_n\}_{n\geqslant 1}$ of real numbers such that*

$$b_n \leqslant a_{nm} \leqslant c_n, \qquad n, m \geqslant 1,$$

and $\displaystyle\sum_{n=1}^{\infty} b_n$ and $\displaystyle\sum_{n=1}^{\infty} c_n$ are convergent. Suppose

$$\lim_{m \to \infty} a_{nm} = \bar{a}_n, \qquad \forall n \geqslant 1.$$

Then $\displaystyle\sum_{n=1}^{\infty} \bar{a}_n$ is convergent and

$$\lim_{m \to \infty} \sum_{n=1}^{\infty} a_{nm} = \sum_{n=1}^{\infty} \bar{a}_n = \sum_{n=1}^{\infty} \lim_{m \to \infty} a_{nm}.$$

(ii) **(Levi's Lemma)** *Suppose*

$$0 \leqslant a_{nm} \leqslant a_{n(m+1)}, \qquad \forall n, m \geqslant 1,$$

and

$$\lim_{m \to \infty} a_{nm} = \bar{a}_n, \qquad n \geqslant 1,$$

Then

$$\lim_{m\to\infty} \sum_{n=1}^{\infty} a_{nm} = \sum_{n=1}^{\infty} \lim_{m\to\infty} a_{nm}.$$

(iii) **(Fatou's Lemma)** *Suppose there exists a sequence* $\{b_n\}_{n\geqslant 1}$ *of real numbers such that*

$$a_{nm} \geqslant b_n, \qquad \forall n, m \geqslant 1,$$

and $\sum_{n=1}^{\infty} b_n$ *is convergent. Then*

$$\sum_{n=1}^{\infty} \lim_{m\to\infty} a_{nm} \leqslant \lim_{m\to\infty} \sum_{n=1}^{\infty} a_{nm}.$$

Proof. (i) For any $\varepsilon > 0$, there exists an $N \geqslant 1$ such that

$$\left| \sum_{n=k}^{\ell} b_n \right| + \left| \sum_{n=k}^{\ell} c_n \right| < \varepsilon, \qquad \forall \ell \geqslant k \geqslant N.$$

Then

$$-\varepsilon < \sum_{n=k}^{\ell} b_n \leqslant \sum_{n=k}^{\ell} a_{nm} \leqslant \sum_{n=k}^{\ell} c_n < \varepsilon, \qquad \ell \geqslant k \geqslant N.$$

Hence, $\sum_{n=1}^{\infty} a_{nm}$ is convergent uniformly in m. Letting $m \to \infty$ in the above, one obtains

$$-\varepsilon \leqslant \sum_{n=k}^{\ell} \bar{a}_n \leqslant -\varepsilon, \qquad \forall \ell \geqslant k \geqslant N.$$

This shows that $\sum_{n=1}^{\infty} \bar{a}_n$ is convergent. Moreover,

$$\left| \sum_{n=1}^{\infty} a_{nm} - \sum_{n=1}^{\infty} \bar{a}_n \right| \leqslant \left| \sum_{n=1}^{N} a_{nm} - \sum_{n=1}^{N} \bar{a}_n \right| + \left| \sum_{k=N+1}^{\infty} a_{nm} \right| + \left| \sum_{n=N+1}^{\infty} \bar{a}_n \right|$$

$$\leqslant \sum_{n=1}^{N} |a_{nm} - \bar{a}_n| + 2\varepsilon.$$

Hence, by sending $m \to \infty$, we obtain our conclusion.

(ii) Since

$$0 \leqslant a_{nm} \leqslant \bar{a}_n, \qquad m, n \geqslant 1,$$

we have

$$\sum_{n=1}^{\infty} a_{nm} \leqslant \sum_{n=1}^{\infty} \bar{a}_n.$$

If we have

$$\lim_{m \to \infty} \sum_{n=1}^{\infty} a_{nm} = \infty,$$

our conclusion is obvious. We now assume that

$$\lim_{m \to \infty} \sum_{n=1}^{\infty} a_{nm} = S < \infty.$$

Then for any $N \geqslant 1$,

$$\sum_{n=1}^{N} a_{nm} \leqslant \sum_{n=1}^{\infty} a_{nm} \leqslant S.$$

Sending $m \to \infty$, one has

$$\sum_{n=1}^{N} \bar{a}_n \leqslant S.$$

Consequently,

$$\sum_{n=1}^{\infty} \bar{a}_n \leqslant S.$$

Then we have the following situation:

$$0 \leqslant a_{nm} \leqslant \bar{a}_n, \qquad \forall n \geqslant 1,$$

with $\sum_{n=1}^{\infty} 0$ and $\sum_{n=1}^{\infty} \bar{a}_n$ convergent. Hence, (i) applies.

(iii) First of all, if the right-hand side is infinite, the conclusion is trivial. Now, let the right-hand side be finite. By

$$b_n \leqslant a_{nm}, \qquad \forall m, n \geqslant 1,$$

one has

$$a_{nm} - b_n \geqslant 0, \qquad \forall m, n \geqslant 1,$$

and

$$\bar{a}_n - b_n \equiv \lim_{m \to \infty} a_{nm} - b_n \geqslant 0, \qquad \forall n \geqslant 1.$$

Hence, for any $N \geqslant 1$,

$$0 \leqslant \sum_{n=1}^{N} (\bar{a}_n - b_n) = \sum_{n=1}^{N} \lim_{m \to \infty} (a_{nm} - b_n)$$

$$\leqslant \lim_{m \to \infty} \sum_{n=1}^{N} (a_{nm} - b_n) \leqslant \lim_{m \to \infty} \sum_{n=1}^{\infty} (a_{nm} - b_n)$$

$$= \lim_{m \to \infty} \sum_{n=1}^{\infty} a_{nm} - \sum_{n=1}^{\infty} b_n < \infty.$$

Then we may let $N \to \infty$ to get

$$\sum_{n=1}^{\infty} (\bar{a}_n - b_n) \leqslant \lim_{m \to \infty} \sum_{n=1}^{\infty} a_{nm} - \sum_{n=1}^{\infty} b_n.$$

This implies that the series on the left-hand side is convergent. Hence,

$$\sum_{n=1}^{\infty} \lim_{m \to \infty} a_{mn} = \sum_{n=1}^{\infty} (\bar{a}_n - b_n) + \sum_{n=1}^{\infty} b_n \leqslant \lim_{m \to \infty} \sum_{n=1}^{\infty} a_{nm}.$$

This proves our conclusion. $\qquad\qquad\square$

The following example shows that if conditions assumed in the above are absent, the corresponding conclusions might fail.

Example 3.7. (i) Let

$$a_{nm} = \delta_{nm} \equiv \begin{cases} 1, & n = m, \\ 0, & n \neq m. \end{cases}$$

Then there exists no c_n controlling a_{nm} from above such that $\sum_{n=1}^{\infty} c_n$ is convergent. In this case,

$$\sum_{n=1}^{\infty} a_{nm} = 1, \quad \forall m \geqslant 1.$$

Whereas,

$$\lim_{m \to \infty} a_{nm} = 0, \quad \forall n \geqslant 1.$$

Thus,

$$\lim_{m \to \infty} \sum_{n=1}^{\infty} a_{nm} = 1 \neq 0 = \sum_{n=1}^{\infty} \lim_{m \to \infty} a_{nm}.$$

(ii) Let

$$a_{nm} = \begin{cases} -m, & n = m, \\ 0, & n \neq m. \end{cases}$$

Then there is no b_n controlling a_{nm} from below such that $\sum_{n=1}^{\infty} b_n$ is convergent. In the case,

$$\lim_{m \to \infty} a_{nm} = 0,$$

whereas,

$$\sum_{n=1}^{\infty} a_{nm} = -m.$$

Hence,

$$\lim_{m \to \infty} \sum_{n=1}^{\infty} a_{nm} = -\infty < 0 = \sum_{n=1}^{\infty} \lim_{m \to \infty} a_{nm}.$$

This example also serves as a counterexample for Levi's Lemma since the monotonicity condition fails.

Exercises

1. Let

$$f(x) = \sum_{n=1}^{\infty} \frac{\sin nx}{n^p}, \qquad x \in \mathbb{R}.$$

Show that $f : \mathbb{R} \to \mathbb{R}$ is continuous for $p > 1$ and is continuously differentiable for $p > 2$.

2. Let

$$f(x) = \sum_{n=0}^{\infty} \frac{(x-1)^n}{n^2}.$$

(i) Show that $f(x)$ is well-defined on $[0, 2]$.

(ii) Find all $x \in [0, 2]$ at which $f(x)$ is continuous, continuously differentiable, respectively.

(iii) Find all $x \in [0, 2]$ at which $f(x)$ is infinite time differentiable.

3. Let $x_k = (x_k^{(n)})_{n \geq 1} \in \ell^p$, i.e.,

$$\|x_k\|_p = \Big(\sum_{n=1}^{\infty} |x_k^{(n)}|^p \Big)^{\frac{1}{p}} < \infty, \qquad k \geq 1.$$

Suppose $\{x_k\}_{k \geqslant 1}$ is Cauchy under the ℓ^p-norm $\| \cdot \|_p$, i.e., for any $\varepsilon > 0$, there exists an $N \geqslant 1$ such that

$$\|x_k - x_m\|_p < \varepsilon, \qquad \forall k, m \geqslant N.$$

Show that there exists an $\bar{x} \equiv (\bar{x}^{(n)})_{n \geqslant 1} \in \ell^p$ such that

$$\lim_{k \to \infty} \|x_k - \bar{x}\|_p = 0.$$

4 More Convergence Criteria

In this section, we will present some more convergence tests for series. A major motivation is to determine if the following type series are convergent:

$$\sum_{n=1}^{\infty} \frac{\sin nx}{n}. \tag{4.1}$$

It is clear that since

$$\left| \frac{\sin nx}{n} \right| \leqslant \frac{1}{n}, \qquad \sum_{n=1}^{\infty} \frac{1}{n} = \infty,$$

Weierstrass M-test fails. We will introduce more powerful tests to determine the convergence of such kind of series.

We first consider series of the form: $\sum a_n b_n$, where $a_n, b_n \in \mathbb{R}$. The following lemma is crucial.

Lemma 4.1. (Abel's Lemma) *Let*

$$a_1 \geqslant a_2 \geqslant \cdots \geqslant a_n, \tag{4.2}$$

and

$$\left| \sum_{i=1}^{k} b_i \right| \leqslant M, \qquad 1 \leqslant k \leqslant n, \tag{4.3}$$

for some $M > 0$. Then

$$\left| \sum_{k=1}^{n} a_k b_k \right| \leqslant M(|a_1| + 2|a_n|). \tag{4.4}$$

Proof. Let

$$B_k = \sum_{i=1}^{k} b_i, \qquad k \geqslant 1.$$

Then (with $B_0 = 0$)

$$b_k = B_k - B_{k-1}, \qquad 1 \leqslant k \leqslant m.$$

Hence,

$$
\begin{aligned}
\sum_{k=1}^{n} a_k b_k &= \sum_{k=1}^{n} a_k \big(B_k - B_{k-1} \big) = \sum_{k=1}^{n} a_k B_k - \sum_{k=1}^{n} a_k B_{k-1} \\
&= \sum_{k=1}^{n} a_k B_k - \sum_{k=1}^{n-1} a_{k+1} B_k = a_n B_n + \sum_{k=1}^{n-1} (a_k - a_{k+1}) B_k.
\end{aligned}
\tag{4.5}
$$

Consequently, by (4.2)–(4.3), we have

$$
\begin{aligned}
\left| \sum_{k=1}^{n} a_k b_k \right| &\leqslant |a_n||B_n| + \sum_{k=1}^{n-1} |a_k - a_{k+1}||B_k| \\
&\leqslant M \left(|a_n| + \sum_{k=1}^{n-1} (a_k - a_{k+1}) \right) = M \big(|a_n| + a_1 - a_n \big) \\
&\leqslant M(|a_1| + 2|a_n|),
\end{aligned}
$$

proving (4.4). $\qquad\qquad\qquad\qquad\qquad\qquad\qquad\qquad\qquad\qquad\qquad$ □

Clearly, if we replace (4.2) by the following:

$$a_1 \leqslant a_2 \leqslant \cdots \leqslant a_n, \tag{4.6}$$

we will still have (4.4).

Theorem 4.2. (Abel's Criterion) *Let $\sum b_n$ be convergent and $\{a_n\}$ be monotone and bounded sequence. Then $\sum a_n b_n$ is convergent.*

Proof. First of all by the convergence of $\sum b_n$, for any $\varepsilon > 0$ there exists an $N \geqslant 1$ such that

$$\left| \sum_{k=n+1}^{n+m} b_k \right| < \varepsilon, \qquad \forall m > 1, \quad n \geqslant N.$$

Then by Lemma 4.1, assuming $|a_n| \leqslant K$ for all $n \geqslant 1$,

$$\left| \sum_{k=n+1}^{n+m} a_k b_k \right| \leqslant \varepsilon \big(|a_{n+1}| + 2|a_{n+m}| \big) \leqslant 3K\varepsilon.$$

Hence, $\sum a_n b_n$ is convergent. $\qquad\qquad\qquad\qquad\qquad\qquad\qquad\qquad$ □

Theorem 4.3. (Dirichlet's Criterion) *Suppose there exists an $M > 0$ such that*

$$\left| \sum_{k=1}^{n} b_k \right| \leqslant M, \qquad \forall n \geqslant 1, \tag{4.7}$$

and a_n converges to 0 monotonically. Then $\sum a_n b_n$ is convergent.

Proof. By the convergence of a_n, for any $\varepsilon > 0$, there exists an $N \geqslant 1$ such that

$$|a_n| < \varepsilon, \qquad \forall n \geqslant N.$$

On the other hand, by (4.7),

$$\left| \sum_{k=1}^{m} b_{n+k} \right| = \left| \sum_{k=1}^{n+m} b_k - \sum_{k=1}^{n} b_k \right| \leqslant \left| \sum_{k=1}^{n+m} b_k \right| + \left| \sum_{k=1}^{n} b_k \right| \leqslant 2M.$$

Hence, by Lemma 4.1, one has

$$\left| \sum_{k=n}^{n+m} a_k b_k \right| = \left| \sum_{k=1}^{m} a_{n+k} b_{n+k} \right| \leqslant 2M \left(|a_{n+1}| + 2|a_{n+m}| \right) < 6M\varepsilon.$$

This gives the convergence. $\qquad\qquad\square$

Example 4.4. Consider series (4.1). We saw that Weierstrass M-test does not work. We now show that for each x,

$$\left| \sum_{k=1}^{n} \sin kx \right| \leqslant M(x),$$

for some $M(x) > 0$. If $x = 2m\pi$, we have

$$\left| \sum_{k=1}^{n} \sin k\pi \right| = 0.$$

Next, we note that

$$\sin \alpha \sin \beta = \frac{1}{2} \Big(\cos(\alpha - \beta) - \cos(\alpha + \beta) \Big).$$

Thus, for $x \neq 2m\pi$,

$$2 \sin \frac{x}{2} \Big(\sin x + \sin 2x + \cdots + \sin nx \Big)$$

$$= \cos \frac{x}{2} - \cos \frac{3x}{2} + \cos \frac{3x}{2} - \cos \frac{5x}{2} + \cdots - \cos \frac{(2n-1)x}{2}$$

$$+ \cos \frac{(2n-1)x}{2} - \cos \frac{(2n+1)x}{2} = \cos \frac{x}{2} - \cos \frac{(2n+1)x}{2}.$$

Consequently,

$$\left| \sum_{k=1}^{n} \sin kx \right| \leqslant \frac{1}{|\sin \frac{x}{2}|} \equiv M(x).$$

Then, by Dirichlet's criterion, we have the (pointwise) convergence of $\sum_{n=1}^{\infty} \dfrac{\sin nx}{n}$. The same idea also applies to

$$\sum_{n=1}^{\infty} \frac{\sin nx}{n^p}, \qquad \sum_{n=0}^{\infty} \frac{\cos nx}{n^p}, \qquad \sum_{n=2}^{\infty} \frac{\sin nx}{(\ln n)^p}, \qquad \sum_{n=2}^{\infty} \frac{\cos nx}{(\ln n)^p}$$

with $p > 0$.

We now look at the following type series: $\sum a_n(x)b_n(x)$.

Theorem 4.5. (Abel's Criterion) *Let* (X, d) *be a metric space and* $\sum b_n(x)$ *be convergent uniformly on* X, *and* $\{a_n(x)\}_{n \geqslant 1}$ *be monotone and uniformly bounded on* X. *Then* $\sum a_n(x)b_n(x)$ *is convergent uniformly.*

Proof. By the uniform convergence of $\sum b_n(x)$, for any $\varepsilon > 0$ there exists an $N \geqslant 1$ such that

$$\left| \sum_{k=n+1}^{n+m} b_k(x) \right| < \varepsilon, \qquad \forall m > 1, \ x \in X, \ n \geqslant N.$$

Then by Lemma 4.1, assuming $|a_n(x)| \leqslant K$ for all $n \geqslant 1$ and x,

$$\left| \sum_{k=n+1}^{n+m} a_k(x)b_k(x) \right| \leqslant \varepsilon \big(|a_{n+1}(x)| + 2|a_{n+m}(x)| \big) \leqslant 3K\varepsilon.$$

Hence, $\sum a_n(x)b_n(x)$ is convergent uniformly. $\qquad \square$

Theorem 4.6. (Dirichlet's Criterion) *Let* (X, d) *be a metric space. Suppose there exists an* $M > 0$ *such that*

$$\left| \sum_{k=1}^{n} b_k(x) \right| \leqslant M, \qquad x \in X, \ \forall n \geqslant 1,$$

and $a_n(x)$ *converges to 0 monotonically and uniformly on* X. *Then the series* $\sum a_n(x)b_n(x)$ *is uniformly convergent.*

Proof. By the convergence of $a_n(x)$, for any $\varepsilon > 0$, there exists an $N \geqslant 1$ such that

$$|a_n(x)| < \varepsilon, \qquad \forall n \geqslant N, \quad \forall x \in X.$$

On the other hand,

$$\left| \sum_{k=1}^{m} b_{n+k}(x) \right| = \left| \sum_{k=1}^{n+m} b_k(x) - \sum_{k=1}^{n} b_k(x) \right|$$

$$\leqslant \left| \sum_{k=1}^{n+m} b_k(x) \right| + \left| \sum_{k=1}^{n} b_k(x) \right| \leqslant 2M.$$

Hence, by Lemma 4.1, one has

$$\left| \sum_{k=1}^{m} a_{n+k}(x)b_{n+k}(x) \right| \leqslant 2M \big(|a_{n+1}(x)| + 2|a_{n+m}(x)| \big) < 6M\varepsilon.$$

This gives the convergence. $\qquad\qquad\qquad\qquad\qquad\qquad\qquad$ $\square$

Exercises

1. Let

$$f(x) = \sum_{n=1}^{\infty} \frac{\sin nx}{n^p}, \qquad x \in (0, 2\pi).$$

Show that $f(\cdot)$ is continuous for any $p > 0$ and is continuously differentiable for any $p > 1$. (Compare with Problem 1 of Section 3.)

2. Discuss the convergence of the series $\displaystyle\sum_{n=1}^{\infty} \frac{\sin nx}{\sqrt{n^4 + x^4}}$.

3. Consider series $\displaystyle\sum_{n=1}^{\infty} \left[\frac{x(x+n)}{n}\right]^n$. Find all values of x such that the series is convergent.

5 Power Series

We now look at a special case of function series:

$$\sum_{n=0}^{\infty} a_n (x - x_0)^n. \tag{5.1}$$

Such a series is called a *power series*.

5.1 *A general consideration*

Definition 5.1. Number $R \geqslant 0$ is called the *radius of convergence* for series (5.1) if

$$\sum_{n=0}^{\infty} a_n (x - x_0)^n \begin{cases} \text{is convergent,} & |x - x_0| < R, \\ \text{is divergent,} & |x - x_0| > R. \end{cases} \tag{5.2}$$

In the above case, we call $(x_0 - R, x_0 + R)$ the *interval of convergence*.

Note that no condition is given for $|x - x_0| = R$ in the above definition.

Proposition 5.2. *The radius of convergence R is given by*

$$R = \left(\varlimsup_{n \to \infty} |a_n|^{\frac{1}{n}} \right)^{-1}. \tag{5.3}$$

Thus, for the case $R > 0$, the following function is well-defined:

$$f(x) = \sum_{n=0}^{\infty} a_n(x - x_0)^n, \qquad x \in (x_0 - R, x_0 + R).$$

Moreover, for any $0 < r < R$, the series $\sum_{n=0}^{\infty} a_n(x - x_0)^n$ is convergent uniformly on $[x_0 - r, x_0 + r]$ to $f(\cdot)$ which is continuous and differentiable with

$$f'(x) = \sum_{n=1}^{\infty} n a_n(x - x_0)^{n-1}, \qquad x \in (x_0 - r, x_0 + r).$$

Further, for any $x_0 - R < a < b < x_0 + R$,

$$\int_a^b \sum_{n=0}^{\infty} a_n(x - x_0)^n dx = \sum_{n=0}^{\infty} a_n \frac{(b - x_0)^{n+1} - (a - x_0)^{n+1}}{n + 1}.$$

Proof. We first show that for any x with $r = |x - x_0| > R$, the series is divergent. In fact, by the definition of R, we can find a subsequence n_k such that for $0 < \varepsilon = \frac{r-R}{Rr}$, there exists a $K \geqslant 1$ so that

$$|a_{n_k}|^{\frac{1}{n_k}} > \frac{1}{R} - \varepsilon = \frac{1}{R} - \frac{r - R}{Rr} = \frac{1}{r}, \qquad k \geqslant K.$$

Then (note $|x - x_0| = r$)

$$|a_{n_k}|^{\frac{1}{n_k}} |x - x_0| > \left(\frac{1}{R} - \varepsilon\right) r = 1, \qquad k \geqslant K.$$

Hence, the series is divergent.

Next, we prove that the series is convergent uniformly on $[x_0 - r, x_0 + r]$ for any $r \in (0, R)$. By the definition of R, for any $0 < \varepsilon < \frac{1}{r} - \frac{1}{R} = \frac{R-r}{Rr}$, when n is large, we have

$$0 < |a_n|^{\frac{1}{n}} r < \left(\frac{1}{R} + \varepsilon\right) r = \frac{(1 + R\varepsilon)r}{R} < 1.$$

Hence, by Weierstrass M-Test, the series is uniformly convergent on $[x_0 - r, x_0 + r]$.

The proof of the rest of the conclusions are straightforward.   □

The following result is concerned with the continuity of the power series at the end point(s) of the interval of convergence.

Theorem 5.3. (Abel) *Let*

$$f(x) = \sum_{n=0}^{\infty} a_n(x - x_0)^n,$$

with the radius of convergence $R \in (0, \infty)$. If $\sum_{n=0}^{\infty} a_n R^n$ is convergent, then $f(\cdot)$ is left-continuous at $x = x_0 + R$, i.e.,

$$\lim_{x \uparrow x_0 + R} \sum_{n=0}^{\infty} a_n (x - x_0)^n = \sum_{n=0}^{\infty} a_n R^n = f(x_0 + R).$$

Likewise, if the series is convergent at $x = x_0 - R$, then $f(\cdot)$ is right-continuous at $x = x_0 - R$.

Before proving the above, let us make a simple remark. If the series $\sum a_n (x - x_0)^n$ is uniformly convergent on $[x_0 - R, x_0 + R]$, then the conclusion follows from the exchange of the order of the limit and summation. Now, we only have uniform convergence on $[x_0 - r, x_0 + r]$ for any $0 < r < R$ and the convergence at $x = x_0 + R$. Therefore, the conclusion is not obvious.

Proof. Without loss of generality, we may assume that $x_0 = 0$ and $R = 1$. Then we need only to prove that

$$\lim_{x \to 1^-} \sum_{n=0}^{\infty} a_n x^n = \sum_{n=0}^{\infty} a_n \equiv S.$$

Let

$$S_N = \left(\sum_{n=0}^{N-1} a_n \right) - S, \qquad N \geqslant 1, \qquad S_0 = -S.$$

Our condition implies

$$\lim_{N \to \infty} S_N = 0, \qquad a_n = S_{n+1} - S_n, \qquad n \geqslant 0.$$

Hence, for any $x \in (-1, 1)$,

$$\sum_{n=0}^{\infty} a_n x^n = \sum_{n=0}^{\infty} (S_{n+1} - S_n) x^n = \sum_{n=0}^{\infty} S_{n+1} x^n - \sum_{n=0}^{\infty} S_n x^n$$

$$= \sum_{n=0}^{\infty} S_{n+1} x^n - \sum_{n=0}^{\infty} S_{n+1} x^{n+1} - S_0 = \sum_{n=0}^{\infty} S_{n+1} (x^n - x^{n+1}) + S.$$

Since $S_n \to 0$, for any $\varepsilon > 0$, there exists an $N \geqslant 1$ such that

$$|S_n| < \varepsilon, \qquad \forall n \geqslant N.$$

For any $x \in [0, 1)$, we have $x^{n+1} \leqslant x^n$. Consequently,

$$\left| \sum_{n=0}^{\infty} a_n x^n - S \right| = \left| \sum_{n=0}^{\infty} S_{n+1}(x^{n+1} - x^n) \right|$$

$$\leqslant \sum_{n=0}^{N} |S_{n+1}|(x^n - x^{n+1}) + \sum_{n=N+1}^{\infty} \varepsilon(x^n - x^{n+1})$$

$$\leqslant \sum_{n=0}^{N} |S_{n+1}|(x^n - x^{n+1}) + \varepsilon x^{N+1}.$$

Now, in the above, we let $x \to 1$. Then

$$\overline{\lim_{x \to 1}} \left| \sum_{n=0}^{\infty} a_n x^n - S \right| \leqslant \varepsilon.$$

Since $\varepsilon > 0$ is arbitrary, we see that

$$\lim_{x \to 1} \sum_{n=0}^{\infty} a_n x^n = S = \sum_{n=0}^{\infty} a_n.$$

This completes the proof. $\qquad\qquad\qquad\qquad\qquad\qquad\qquad$ □

Definition 5.4. Let $E \subseteq \mathbb{R}$ and $f : E \to \mathbb{R}$, x_0 be an interior point of E. We say that $f(\cdot)$ is *real analytic* at $x = x_0$ if there exists an $r > 0$ such that

$$f(x) = \sum_{n=0}^{\infty} a_n(x - x_0)^n, \qquad \forall x \in (x_0 - r, x_0 + r) \subseteq E.$$

If E is open, and $f(\cdot)$ is real analytic at every point of E, we say that $f(\cdot)$ is real analytic on E.

Proposition 5.5. *Let $f : E \to \mathbb{R}$ be real analytic at x_0 having the following power series representation:*

$$f(x) = \sum_{n=0}^{\infty} a_n(x - x_0)^n, \qquad x \in (x_0 - r, x_0 + r).$$

Then $f(\cdot)$ is infinitely differentiable at x_0, i.e., for any $k \geqslant 1$, $f^{(k)}(x_0)$ exists. Moreover,

$$a_n = \frac{f^{(n)}(x_0)}{n!}, \qquad n \geqslant 0.$$

Consequently, $f(\cdot)$ admits the following Taylor expansion:

$$f(x) = \sum_{n=0}^{\infty} \frac{f^{(n)}(x_0)}{n!}(x - x_0)^n, \qquad x \in (x_0 - r, x_0 + r),$$

and if $f(\cdot)$ admits the following power series representation:

$$f(x) = \sum_{n=0}^{\infty} b_n (x - x_0)^n, \qquad x \in (x_0 - r, x_0 + r),$$

then it is necessary that

$$a_n = b_n, \qquad \forall n \geqslant 1.$$

Proof. Without loss of generality, we let $x_0 = 0$. Then by definition, there exists a power series such that

$$f(x) = \sum_{n=0}^{\infty} a_n x^n.$$

By the uniform convergence of the power series in any compact subsets of $(-r, r)$, one can obtain

$$f^{(k)}(x) = \sum_{n=0}^{\infty} a_{n+k} \frac{(n+k)!}{n!} x^n, \qquad x \in (-r, r).$$

This proves the differentiability of $f(\cdot)$. The rest conclusions are clear.

$\square$

Example 5.6. Let

$$f(x) = \begin{cases} 0, & x \leqslant 0, \\ e^{-\frac{1}{x^2}}, & x > 0. \end{cases}$$

Then

$$f^{(k)}(x) = \begin{cases} 0, & x \leqslant 0, \\ R_k(x) e^{-\frac{1}{x^2}}, & x > 0, \end{cases}$$

with $R_k(\cdot)$ being a rational function. Hence, in particular,

$$f^{(k)}(0) = 0, \qquad k \geqslant 0.$$

Then for any $\delta > 0$,

$$\sum_{n=0}^{\infty} \frac{f^{(n)}(0)}{n!} x^n \equiv 0 \neq f(x), \qquad x \in (0, \delta).$$

This means that $f(\cdot)$ is infinitely differentiable at $x = 0$, but not real analytic.

Next result is about the product of two power series.

Proposition 5.7. Let $f, g : (x_0 - r, x_0 + r) \to \mathbb{R}$ be real analytic with power series representations:

$$f(x) = \sum_{n=0}^{\infty} a_n (x - x_0)^n, \qquad g(x) = \sum_{n=0}^{\infty} b_n (x - x_0)^n,$$

which are convergent in $(x_0 - r, x_0 + r)$, for some $r > 0$. Then the product $f(\cdot)g(\cdot)$ is also real analytic on $(x_0 - r, x_0 + r)$ with power series expansion:

$$f(x)g(x) = \sum_{k=0}^{\infty} c_k (x - x_0)^k,$$

where

$$c_k = \sum_{m=0}^{k} a_m b_{k-m}, \quad k \geqslant 0.$$

Proof. Without loss of generality, we let $x_0 = 0$. For any $\delta \in (0, r)$, consider $x \in [-r + \delta, r - \delta]$. Since the power series for $f(\cdot)$ and $g(\cdot)$ are uniformly and absolutely convergent on $[-r + \delta, +r - \delta]$, for any $\varepsilon > 0$, there exists an $N \geqslant 1$ such that

$$\sum_{n=N+1}^{\infty} |a_n| |x|^n + \sum_{n=N+1}^{\infty} |b_n| |x|^n < \varepsilon, \quad |x| \leqslant r - \delta.$$

Now,

$$\left(\sum_{n=0}^{N} a_n x^n \right) \left(\sum_{m=0}^{M} b_m x^m \right) = \sum_{n=0}^{N} \sum_{m=0}^{M} a_n b_m x^{n+m} = \sum_{k=0}^{N+M} c_k x^k.$$

Further,

$$\left| \sum_{k=0}^{N+M} c_k x^k - f(x)g(x) \right|$$

$$= \left| \sum_{k=0}^{N+M} c_k x^k - \left(\sum_{n=0}^{N} a_n x^n + \sum_{n=N+1}^{\infty} a_n x^n \right) \left(\sum_{n=0}^{M} b_n x^n + \sum_{n=M+1}^{\infty} b_n x^n \right) \right|$$

$$\leqslant \left| \sum_{n=0}^{N} a_n x^n \right| \left(\sum_{n=M+1}^{\infty} |b_n| |x|^n \right) + \left| \sum_{n=0}^{M} b_n x^n \right| \left(\sum_{n=N+1}^{\infty} |a_n| |x|^n \right)$$

$$+ \left(\sum_{n=N+1}^{\infty} |a_n| |x|^n \right) \left(\sum_{n=M+1}^{\infty} |b_n| |x|^n \right)$$

$$\leqslant \left(|f(x)| + \varepsilon \right) \varepsilon + \left(|g(x)| + \varepsilon \right) \varepsilon + \varepsilon^2$$

$$\leqslant \varepsilon \left[\max_{|x| \leqslant r - \delta} \left(|f(x)| + |g(x)| \right) \right] + 3\varepsilon^2.$$

Hence,

$$f(x)g(x) = \sum_{k=0}^{\infty} c_k x^k,$$

proving our conclusion. □

5.2 *Exponential and logarithmic functions*

In this subsection, we construct exponential and logarithmic functions by power series.

Example 5.8. Let

$$E(x) = \sum_{n=0}^{\infty} \frac{x^n}{n!}, \tag{5.4}$$

First of all, *Stirling's formula* says

$$\lim_{n \to \infty} \frac{n! e^n}{n^n \sqrt{2n\pi}} = 1. \tag{5.5}$$

Applying the above, we calculate

$$\varlimsup_{n \to \infty} \frac{1}{(n!)^{\frac{1}{n}}} = \varlimsup_{n \to \infty} \left[\left(\frac{e}{n} \right)^n \right]^{\frac{1}{n}} = 0.$$

Hence, the radius of convergence for the series (5.4) is $R = \infty$. In other words, the series converges for all $x \in \mathbb{R}$. Further, it is clear that the convergence is uniform in any compact sets, and the series of formal term-by-term derivatives has the same form and therefore converges uniformly in any compact sets. Hence, we have

$$E'(x) = E(x), \qquad \forall x \in \mathbb{R}.$$

Likewise, we can integrate term-by-term to get

$$\int_a^b E(x)dx = E(b) - E(a).$$

Further, for any $x, y \in \mathbb{R}$,

$$E(x)E(y) = \left(\sum_{n=0}^{\infty} \frac{x^n}{n!} \right) \left(\sum_{m=0}^{\infty} \frac{y^m}{m!} \right) = \sum_{k=0}^{\infty} \sum_{n+m=k} \frac{x^n y^m}{n!m!}$$

$$= \sum_{k=0}^{\infty} \sum_{n=0}^{k} \frac{x^n y^{k-n}}{n!(k-n)!} = \sum_{k=0}^{\infty} \frac{(x+y)^k}{k!} = E(x+y).$$

Thus,

$$E(x) = \left[E\left(\frac{x}{2}\right)\right]^2 \geqslant 0.$$

We also have

$$1 = E(0) = E(x - x) = E(x)E(-x),$$

leading to

$$\begin{cases} E(x) > 0, \\ E(-x) = \dfrac{1}{E(x)}, \end{cases} \quad \forall x \in \mathbb{R}.$$

Define the *Euler's number* by the following:

$$e = E(1) = \sum_{n=0}^{\infty} \frac{1}{n!},$$

Then, by induction, for any integer $m > 1$,

$$E(m) = E(1)^m = e^m,$$

and

$$e = E\left(\frac{m}{m}\right) = E\left(\frac{1}{m}\right)^m,$$

which leads to

$$E\left(\frac{1}{m}\right) = e^{\frac{1}{m}}.$$

Consequently, for any rational number $r = \frac{p}{q} > 0$,

$$E\left(\frac{p}{q}\right) = e^{\frac{p}{q}}, \qquad E\left(-\frac{p}{q}\right) = \frac{1}{E\left(\frac{p}{q}\right)} = e^{-\frac{p}{q}}.$$

Finally, by the continuity of the map $x \mapsto E(x)$, we obtain that

$$E(x) = e^x, \qquad \forall x \in \mathbb{R}.$$

This function is called the *exponential function*.

Example 5.9. Consider the following series:

$$\varphi(x) = \sum_{n=1}^{\infty} \frac{(-1)^{n+1}}{n}(x - 1)^n.$$

Since

$$\varlimsup_{n \to \infty} \left(\frac{1}{n}\right)^{\frac{1}{n}} = 1,$$

we obtain that the radius of convergence is equal to 1. Thus, the above series is convergent for $x \in (0, 2)$, and the function $\varphi(\cdot)$ is real analytic on $(0, 2)$. Further, in this interval,

$$\varphi'(x) = \sum_{n=1}^{\infty} (-1)^{n-1}(x-1)^{n-1} = \sum_{n=0}^{\infty} (1-x)^n = \frac{1}{1-(1-x)} = \frac{1}{x}.$$

Consequently, by the chain rule,

$$[\varphi(e^x)]' = \varphi'(e^x)e^x = \frac{1}{e^x}e^x = 1, \qquad \varphi(e^0) = \varphi(1) = 0.$$

Hence,

$$\varphi(e^x) = x, \qquad x \in (0, 2).$$

This means that $\varphi(\cdot)$ is an inverse function of e^x. We denote

$$\varphi(x) = \ln x, \qquad x \in (0, 2).$$

We can extend it to $(0, \infty)$ as an inverse function of $x \mapsto e^x$. This function is called the *logarithmic function*.

5.3   The sine and cosine functions

In this subsection, we construct sine and cosine functions by power series. First of all, we recall the set $\mathbb{C}$ of all complex numbers:

$$\mathbb{C} = \{a + ib \mid a, b \in \mathbb{R}\}, \qquad i^2 = -1,$$

with the usual rules of addition, multiplication, etc. Also, the conjugate of $a + ib$ is defined by

$$\overline{a + ib} = a - ib$$

and the norm of $a + ib$ is given by

$$|a + ib| = \sqrt{a^2 + b^2}.$$

Next, we extend the exponential function e^x from $\mathbb{R}$ to $\mathbb{C}$:

$$e^z = \sum_{n=0}^{\infty} \frac{z^n}{n!}, \qquad z \in \mathbb{C}.$$

It is clear that the series is convergent for all $z \in \mathbb{C}$. Now, we define:

$$C(x) = \frac{e^{ix} + e^{-ix}}{2} = \frac{1}{2} \sum_{n=0}^{\infty} \frac{(ix)^n + (-ix)^n}{n!} = \sum_{m=0}^{\infty} \frac{(-1)^m x^{2m}}{(2m)!},$$

$$S(x) = \frac{e^{ix} - e^{-ix}}{2i} = \frac{1}{2i} \sum_{n=0}^{\infty} \frac{(ix)^n - (-ix)^n}{n!} = \sum_{m=0}^{\infty} \frac{(-1)^m x^{2m+1}}{(2m+1)!},$$

$$x \in \mathbb{R}.$$

Then $C, S : \mathbb{R} \to \mathbb{R}$ are real analytic on $\mathbb{R}$, and the following holds:

$$e^{ix} = C(x) + iS(x), \qquad \forall x \in \mathbb{R}. \qquad (5.6)$$

It is seen from the definition that

$$\begin{cases} C(x)^2 + S(x)^2 = |e^{ix}|^2 = e^{ix}e^{-ix} = e^0 = 1, & \forall x \in \mathbb{R}, \\ C'(x) = -S(x), \quad S'(x) = C(x), & \forall x \in \mathbb{R}, \\ C(0) = 1, \quad S(0) = 0. \end{cases} \qquad (5.7)$$

We claim that there exists an $\bar{x} > 0$ such that

$$C(\bar{x}) = 0.$$

If this is not the case, then from $C(0) = 1 > 0$, we must have

$$C(x) > 0, \qquad \forall x \geqslant 0.$$

Consequently,

$$S'(x) = C(x) > 0, \qquad \forall x \in \mathbb{R}.$$

Hence, $S(\cdot)$ is strictly increasing. By $S(0) = 0$, we must have

$$S(x) > 0, \qquad \forall x > 0.$$

Now, for any $0 < x < y$, one has

$$S(x)(y - x) < \int_x^y S(t)dt = C(x) - C(y) \leqslant 1.$$

This will lead to a contradiction since $S(x) > 0$ and we may let y large. Now, we let $\bar{x} > 0$ be the smallest positive number such that $C(\bar{x}) = 0$. Denote

$$\pi = 2\bar{x}.$$

Note that, at the moment, π is just a notation. Then by (5.7),

$$C\left(\frac{\pi}{2}\right) = C(\bar{x}) = 0, \qquad S\left(\frac{\pi}{2}\right) = S(\bar{x}) = \pm 1. \qquad (5.8)$$

Since

$$C(x) > 0, \qquad x \in \left[0, \frac{\pi}{2}\right) = [0, \bar{x}), \qquad (5.9)$$

by $S(0) = 0$ and $S'(x) = C(x)$, we see that

$$S(x) > 0, \qquad x \in \left(0, \frac{\pi}{2}\right], \qquad (5.10)$$

which leads to

$$S\left(\frac{\pi}{2}\right) = 1, \qquad 0 < S(x) < 1, \quad x \in (0, \frac{\pi}{2}). \qquad (5.11)$$

Thus,

$$e^{\frac{\pi i}{2}} = C\left(\frac{\pi}{2}\right) + iS\left(\frac{\pi}{2}\right) = i.$$

Consequently,

$$e^{\pi i} + 1 = 0, \qquad e^{2\pi i} = 1.$$

Therefore,

$$e^{z+2\pi i} = e^z, \qquad \forall z \in \mathbb{C}. \tag{5.12}$$

Then

$$C(x + 2\pi) + iS(x + 2\pi) = e^{ix+2\pi i} = e^{ix} = C(x) + iS(x), \qquad \forall x \in \mathbb{R},$$

which yields

$$C(x + 2\pi) = C(x), \qquad S(x + 2\pi) = S(x), \qquad \forall x \in \mathbb{R}. \tag{5.13}$$

Next, we claim that

$$e^{it} \neq 1, \qquad \forall 0 < t < 2\pi. \tag{5.14}$$

Suppose otherwise, there exists a $t \in (0, 2\pi)$ such that

$$e^{it} = 1.$$

We write

$$e^{\frac{t}{4}i} = C\left(\frac{t}{4}\right) + iS\left(\frac{t}{4}\right) = x + iy,$$

with $x, y \in \mathbb{R}$. Then

$$1 = e^{it} = (x + iy)^4 = x^4 + 4x^3(iy) + 6x^2(iy)^2 + 4x(iy)^3 + (iy)^4$$
$$= x^4 - 6x^2y^2 + y^4 + 4ixy(x^2 - y^2).$$

Hence, it is necessary that

$$xy(x^2 - y^2) = 0.$$

On the other hand, since $\frac{t}{4} \in (0, \frac{\pi}{2})$, by (5.9) and (5.10), we have

$$0 < x, y < 1.$$

Thus, we must have

$$x^2 - y^2 = 0, \qquad x^2 + y^2 = 1,$$

which implies

$$x^2 = y^2 = \frac{1}{2}.$$

As a result, one has

$$1 = x^4 - 6x^2y^2 + y^4 = \frac{1}{4} - \frac{6}{4} + \frac{1}{4} = -1,$$

a contradiction. This proves our claim (5.14).

Further, we show that for any $z \in \mathbb{C}$ with $|z| = 1$, there exists a unique $t \in [0, 2\pi)$ such that

$$e^{it} = z. \tag{5.15}$$

In fact, for any $0 \leqslant t_1 < t_2 < 2\pi$,

$$1 \neq e^{i(t_2 - t_1)} = e^{it_2} e^{-it_1},$$

which leads to

$$e^{it_1} \neq e^{it_2}.$$

This gives the uniqueness. For the existence, we write

$$z = x + iy, \qquad x, y \in \mathbb{R}, \qquad x^2 + y^2 = |z|^2 = 1.$$

Case 1. $x, y \geqslant 0$. On $[0, \frac{\pi}{2}]$, $C(\cdot)$ decreases from 1 to zero (see (5.7) and (5.8)). Hence, for $x \in [0, 1]$, there exists a $t \in [0, \frac{\pi}{2}]$ such that

$$C(t) = x.$$

Then by (5.7) and (5.10), we obtain

$$S(t) = \sqrt{1 - C(t)^2} = y,$$

leading to

$$e^{it} = x + iy = z,$$

giving the existence for this case (see (5.6)).

Case 2. $x < 0$ *and* $y \geqslant 0$. In this case,

$$-iz = y + i(-x), \qquad y, -x \geqslant 0.$$

Hence, there exists a $t \in [0, \frac{\pi}{2}]$ such that

$$e^{it} = -iz.$$

Then

$$z = ie^{it} = e^{\frac{\pi i}{2}} e^{it} = e^{i(t + \frac{\pi}{2})}, \qquad t + \frac{\pi}{2} \in [0, \pi).$$

Case 3. $y < 0$. In this case,

$$-z = -x - iy = -x + i(-y), \qquad -y > 0.$$

Hence, by Cases 1 and 2, there exists a $t \in [0, \pi)$ such that
$$-z = e^{it}.$$
Then
$$z = -e^{it} = e^{i(t+\pi)}.$$
This completes the proof of (5.15).

Finally, we identify π. To this end, we define a curve
$$\gamma(t) = e^{it}, \qquad t \in [0, 2\pi],$$
which is the unit circle in $\mathbb{C}$. The arc length is given by
$$\int_0^{2\pi} |\gamma'(t)| dt = \int_0^{2\pi} |ie^{it}| dt = \int_0^{2\pi} dt = 2\pi.$$
This coincides the usual definition of π.

We denote
$$\cos x = C(x), \qquad \sin x = S(x), \qquad \forall x \in \mathbb{R},$$
and call them *cosine function* and *sine function*, respectively. These are *trigonometric functions*.

Definition 5.10. Let $T > 0$. A function $f : \mathbb{R} \to \mathbb{C}$ is *periodic* with *period* T, or *T-periodic*, if
$$f(x + T) = f(x), \qquad x \in \mathbb{R}.$$

From the above, we see that $\cos x$ and $\sin x$ are periodic functions with period 2π.

Exercises

1. Find a rational function $f(x)$ such that
$$f(x) = \sum_{n=1}^{\infty} n^2 x^n, \qquad |x| < 1.$$

2. Consider series $\sum_{n=0}^{\infty} a_n x^n$. Find a rational function $f(x)$ such that
$$f(x) = \sum_{n=0}^{\infty} a_n x^n, \qquad |x| < R,$$
where R is the radius of convergence for the series, to be determined, for each of the following cases:

(i) Let $a_n = 2n + 1$, $n \geqslant 0$.

(ii) Let $a_{n+1} = a_n + a_{n-1}$ for $n \geqslant 1$, $a_0 = a_1 = 1$.

(iii) Let $a_{n+1} = \alpha a_n + \beta a_{n-1}$ for $n \geqslant 1$, and $a_0, a_1, \alpha, \beta \in \mathbb{R}$ are given.

6 Fourier Series

We first introduce the following definition.

Definition 6.1. The following is called a *trigonometric polynomial*:

$$f(x) = a_0 + \sum_{n=1}^{N}(a_n \cos nx + b_n \sin nx), \qquad x \in \mathbb{R}, \tag{6.1}$$

where $a_n, b_n \in \mathbb{C}$. If all coefficients $a_n, b_n \in \mathbb{R}$, the above polynomial $f(\cdot)$ is said to be *real*.

Note that

$$\begin{pmatrix} \cos nx \\ \sin nx \end{pmatrix} = \frac{1}{2}\begin{pmatrix} 1 & 1 \\ -i & i \end{pmatrix}\begin{pmatrix} e^{inx} \\ e^{-inx} \end{pmatrix}, \quad \begin{pmatrix} e^{inx} \\ e^{-inx} \end{pmatrix} = \begin{pmatrix} 1 & i \\ 1 & -i \end{pmatrix}\begin{pmatrix} \cos nx \\ \sin nx \end{pmatrix}. \tag{6.2}$$

Therefore, the above (6.1) can also be equivalently written as

$$f(x) = \sum_{n=-N}^{N} c_n e^{inx}, \qquad x \in \mathbb{R}, \tag{6.3}$$

for some $c_n \in \mathbb{C}$. It is clear that for $f(\cdot)$ of form (6.3) to be real-valued for all x if and only if

$$c_n = \bar{c}_n, \qquad \forall n = 0, \pm 1, \pm 2, \cdots, \pm N.$$

Note that the domain of the above function is $\mathbb{R}$ (with possibly complex valued).

In this section, we will present a theory that represent functions in terms of the following type series:

$$\sum_{n=-\infty}^{\infty} c_n e^{inx},$$

or

$$\sum_{n=0}^{\infty} a_n \cos nx + \sum_{n=1}^{\infty} b_n \sin nx.$$

We call the above type series *trigonometric series*, and when the coefficient c_n, or a_n and b_n are properly obtained, they are called *Fourier series*.

6.1 *Inner product*

Let $C([a, b]; \mathbb{C})$ be the set of all continuous functions $f : [a, b] \to \mathbb{C}$.

Definition 6.2. For any $f(\cdot), g(\cdot) \in C([a, b]; \mathbb{C})$, the following is called an *inner product* of $f(\cdot)$ and $g(\cdot)$:

$$\langle f, g \rangle = \int_a^b f(x)\overline{g(x)}dx, \qquad \forall f, g \in C([a, b]; \mathbb{C}).$$

Proposition 6.3. *The inner product* $\langle \cdot, \cdot \rangle$ *satisfies the following:*

(i) **(Hermitian property)** *For any* $f(\cdot), g(\cdot) \in C([a, b]; \mathbb{C})$,

$$\langle f, g \rangle = \overline{\langle g, f \rangle}.$$

(ii) **(Positivity)** *For any* $f(\cdot) \in C([a, b]; \mathbb{C})$,

$$\langle f, f \rangle \geqslant 0, \qquad \text{and} \qquad \langle f, f \rangle = 0 \quad \Longleftrightarrow \quad f(\cdot) = 0.$$

(iii) **(Linearity)** *For any* $f(\cdot), g(\cdot), h(\cdot) \in C([a, b]; \mathbb{C})$ *and* $\alpha, \beta \in \mathbb{C}$, *it holds*

$$\langle \alpha f + \beta g, h \rangle = \alpha \langle f, h \rangle + \beta \langle g, h \rangle,$$
$$\langle f, \alpha g + \beta h \rangle = \bar{\alpha} \langle f, g \rangle + \bar{\beta} \langle f, h \rangle.$$

The proof is straightforward.

We now define

$$\|f\|_2 = \left(\int_a^b |f(x)|^2 dx \right)^{\frac{1}{2}}, \qquad \forall f(\cdot) \in C([a, b]; \mathbb{C}).$$

This is called the L^2-*norm* on $C([a, b]; \mathbb{C})$, for which one has the following result whose proof is straightforward as well.

Proposition 6.4. *The* L^2-*norm* $\| \cdot \|_2$ *satisfies the following:*

(i) **(Non-degeneracy)** *For any* $f(\cdot) \in C([a, b]; \mathbb{C})$,

$$\|f(\cdot)\|_2 \geqslant 0, \qquad \text{and} \qquad \|f\|_2 = 0 \quad \Longleftrightarrow \quad f(\cdot) = 0.$$

(ii) **(Cauchy-Schwarz Inequality)** *For any* $f(\cdot), g(\cdot) \in C([a, b]; \mathbb{C})$,

$$|\langle f, g \rangle| \leqslant \|f\|_2 \|g\|_2.$$

(iii) **(Triangle Inequality)** *For any* $f(\cdot), g(\cdot) \in C([a, b]; \mathbb{C})$,

$$\|f + g\|_2 \leqslant \|f\|_2 + \|g\|_2.$$

(iv) **(Pythagorean Theorem)** *For any* $f(\cdot), g(\cdot) \in C([a,b]; \mathbb{C})$, *with* $\langle f, g \rangle = 0$ *(which is called orthogonal)*

$$\|f + g\|_2^2 = \|f\|_2^2 + \|g\|_2^2.$$

(v) **(Positive Homogeneity)** *For any* $f(\cdot) \in C([a,b]; \mathbb{C})$ *and* $c \in \mathbb{C}$,

$$\|cf\|_2 = |c| \, \|f\|_2.$$

We define

$$d_2(f,g) = \|f - g\|_2 = \left(\int_a^b |f(x) - g(x)|^2 dx \right)^{\frac{1}{2}},$$

$$\forall f(\cdot), g(\cdot) \in C([a,b]; \mathbb{C}).$$

Then $d_2(\cdot, \cdot)$ is a metric on $C([a,b]; \mathbb{C})$.

Warning: $C([a,b]; \mathbb{C})$ is not complete under the metric $d_2(\cdot, \cdot)$. The completion of $C([a,b]; \mathbb{C})$ under L^2-norm (with the integral being Lebesgue integral) is a *Hilbert space*, and denoted by $L^2(a, b; \mathbb{C})$.

Definition 6.5. Let $\phi_n : [a,b] \to \mathbb{C}$ be a sequence of continuous complex functions. Then $\{\phi_n\}$ is called an *orthogonal system* of $C([a,b]; \mathbb{C})$ if

$$\langle \phi_m, \phi_n \rangle = \int_a^b \phi_m(x)\overline{\phi_n(x)}dx = 0, \qquad m \neq n.$$

If, in addition,

$$\langle \phi_n, \phi_n \rangle = \int_a^b |\phi_n(x)|^2 dx = 1, \qquad \forall n,$$

we call $\{\phi_n\}$ an *orthonormal system*.

The following example will give us the orthogonal system that we will use below.

Example 6.6. (i) Consider $C([-\pi, \pi]; \mathbb{C})$.

$$\langle 1, e^{ix} \rangle = \int_{-\pi}^{\pi} 1 e^{-ix} dx = -\frac{1}{i} e^{-ix} \Big|_{-\pi}^{\pi} = 0.$$

For $m, n \in \mathbb{Z}$ with $m \neq n$, we have

$$\langle e^{imx}, e^{inx} \rangle = \int_{-\pi}^{\pi} e^{i(m-n)x} dx = \frac{1}{i(m-n)} e^{i(m-n)x} \Big|_{-\pi}^{\pi} = 0,$$

and

$$\langle e^{imx}, e^{imx} \rangle = \int_{-\pi}^{\pi} e^{i(m-m)x} dx = \int_{-\pi}^{\pi} dx = 2\pi.$$

Hence, $\{e^{inx} \mid n \in \mathbb{Z}\}$ is an orthogonal system of $C([-\pi, \pi]; \mathbb{C})$. If we define

$$e_n(x) = \frac{1}{\sqrt{2\pi}} e^{inx}, \qquad x \in [-\pi, \pi], \quad n \in \mathbb{Z}, \tag{6.4}$$

then $\{e_n(\cdot) \mid n \in \mathbb{Z}\}$ is an orthonormal system of $C([-\pi, \pi]; \mathbb{C})$.

(ii) Consider $C([-\pi, \pi]; \mathbb{R})$. If $m, n \in \mathbb{N} \cup \{0\}$ with $m \neq n$, then

$$\langle \cos mx, \cos nx \rangle = \int_{-\pi}^{\pi} \cos mx \cos nx\, dx$$

$$= \frac{1}{2} \int_{-\pi}^{\pi} \Big(\cos(m+n)x + \cos(m-n)x \Big)\, dx = 0,$$

and

$$\langle \cos mx, \cos mx \rangle = \int_{-\pi}^{\pi} \cos^2 mx\, dx = \int_{-\pi}^{\pi} \frac{1 + \cos 2mx}{2}\, dx = \pi.$$

If $m, n \in \mathbb{N}$ with $m \neq n$, then

$$\langle \sin mx, \sin nx \rangle = \int_{-\pi}^{\pi} \sin mx \sin nx\, dx$$

$$= \frac{1}{2} \int_{-\pi}^{\pi} \Big(\cos(m-n)x - \cos(m+n)x \Big)\, dx = 0,$$

and

$$\langle \sin mx, \sin mx \rangle = \int_{-\pi}^{\pi} \sin^2 mx\, dx = \int_{-\pi}^{\pi} \frac{1 - \cos 2mx}{2}\, dx = \pi.$$

Finally, for any $n \geqslant 1$, $m \geqslant 0$,

$$\langle \sin nx, \cos mx \rangle = \int_{-\pi}^{\pi} \sin nx \cos mx\, dx$$

$$= \frac{1}{2} \int_{-\pi}^{\pi} \Big(\sin(n+m)x + \sin(n-m)x \Big)\, dx = 0.$$

Therefore, $\{1, \cos nx, \sin nx \mid n \in \mathbb{N}\}$ is an orthogonal system of $C([-\pi, \pi]; \mathbb{R})$. Further,

$$\frac{1}{\sqrt{2\pi}}, \quad \frac{\cos nx}{\sqrt{\pi}}, \quad \frac{\sin nx}{\sqrt{\pi}}, \quad n \geqslant 1, \tag{6.5}$$

form an orthonormal system of $C([-\pi, \pi]; \mathbb{R})$.

6.2   *Trigonometric polynomial approximation*

Recall from Chapter 5, Theorem 3.6, we know that any $f(\cdot) \in C([a, b]; \mathbb{R})$ can be uniformly approximated by polynomials. The following gives another type of approximation of continuous functions.

Theorem 6.7. (Weierstrass) *Let $f(\cdot) \in C([-\pi, \pi]; \mathbb{C})$. Then there exists a sequence of trigonometric polynomial $T_n(\cdot)$ such that*

$$\lim_{n \to \infty} \|f(\cdot) - T_n(\cdot)\|_\infty = 0.$$

Proof. Let

$$\Phi_N(x) = \frac{1}{\sqrt{2\pi}} \sum_{n=-(N-1)}^{N-1} \left(1 - \frac{|n|}{N}\right) e_n(x).$$

Note that

$$\sqrt{2\pi} \left| \sum_{n=0}^{N-1} e_n \right|^2 = \frac{1}{\sqrt{2\pi}} \sum_{n,m=0}^{N-1} e^{i(n-m)x} = \sum_{n,m=0}^{N-1} e_{n-m}$$

$$= Ne_0 + (N-1)e_1 + (N-2)e_2 + \cdots + 2e_{N-2} + e_{N-1}$$

$$+ (N-1)e_{-1} + (N-2)e_{-2} + \cdots + 2e_{-(N-2)} + e_{-(N-1)}$$

$$= \sum_{n=0}^{N-1} (N-n)e_n + \sum_{n=1}^{N-1} (N-n)e_{-n} = \sum_{n=0}^{N-1} (N-n)e_n + \sum_{n=-1}^{-(N-1)} (N-|n|)e_n$$

$$= \sum_{n=-(N-1)}^{N-1} (N-|n|)e_n = \sqrt{2\pi}N\Phi_N.$$

On the other hand, when $0 < |x| \leqslant \pi$,

$$\sum_{n=0}^{N-1} e_n = \frac{1}{\sqrt{2\pi}} \sum_{n=0}^{N-1} e^{inx} = \frac{1}{\sqrt{2\pi}} \frac{e^{iNx} - 1}{e^{ix} - 1} = \frac{1}{\sqrt{2\pi}} e^{i\frac{(N-1)x}{2}} \frac{\sin \frac{Nx}{2}}{\sin \frac{x}{2}}.$$

Hence,

$$0 \leqslant \Phi_N(x) = \frac{1}{N} \left| \sum_{n=0}^{N-1} e_n(x) \right|^2 = \frac{1}{2N\pi} \frac{\sin^2 \frac{Nx}{2}}{\sin^2 \frac{x}{2}} \leqslant \frac{1}{2\pi \sin^2 \frac{x}{2}}. \tag{6.6}$$

Also,

$$\Phi_N(0) = \frac{N}{2\pi}.$$

Further,

$$\int_{-\pi}^{\pi} \Phi_N(x)dx = \sum_{n=-(N-1)}^{N-1} \left(1 - \frac{|n|}{N}\right) \frac{1}{2\pi} \int_{-\pi}^{\pi} e^{inx} dx = 1.$$

Next, by the continuity of $f(\cdot)$, we have

$$|f(x)| \leqslant M, \qquad \forall x \in [-\pi, \pi],$$

and for any $\varepsilon > 0$, there exists a $\delta > 0$ such that

$$|f(x) - f(y)| \leqslant \varepsilon, \qquad |x - y| \leqslant \delta.$$

By (6.6), for the above $\delta > 0$, there exists an $N \geqslant 1$ such that

$$0 \leqslant \Phi_N(x) \leqslant \frac{1}{2N\pi \sin^2 \frac{\delta}{2}} \leqslant \varepsilon, \qquad \forall \delta \leqslant |x| \leqslant \pi.$$

Now, we extend $f(\cdot)$ periodically outside of $[-\pi, \pi]$, and define

$$
\begin{aligned}
(f * \Phi_N)(x) &= \int_{-\pi}^{\pi} f(t)\Phi_N(x - t)dt = \int_{-\pi+x}^{x+\pi} f(x - \tau)\Phi_N(\tau)d\tau \\
&= \left(\int_{-\pi+x}^{\pi} + \int_{\pi}^{\pi+x} \right) f(x - \tau)\Phi_N(\tau)d\tau \\
&= \left(\int_{-\pi+x}^{\pi} + \int_{-\pi}^{-\pi+x} \right) f(x - \tau)\Phi_N(\tau)d\tau \\
&= \int_{-\pi}^{\pi} f(x - \tau)\Phi_N(\tau)d\tau, \quad x \in \mathbb{R}.
\end{aligned}
$$

Clearly, the above is a trigonometric polynomial. We now estimate

$$
\begin{aligned}
|f(x) - f * \Phi_N(x)| &= \left| f(x) - \int_{-\pi}^{\pi} f(x - t)\Phi_N(t)dt \right| \\
&= \left| \int_{-\pi}^{\pi} \Big(f(x) - f(x - y) \Big) \Phi_N(y)dy \right| \\
&\leqslant \int_{-\pi}^{\pi} |f(x) - f(x - y)| \Phi_N(y)dy \\
&\leqslant \int_{-\pi}^{-\delta} 2M\varepsilon dy + \int_{-\delta}^{\delta} \varepsilon \Phi_N(y)dy + \int_{\delta}^{\pi} 2M\varepsilon dy \\
&\leqslant 4M\varepsilon(\pi - \delta) + \varepsilon \leqslant (4M\pi + 1)\varepsilon.
\end{aligned}
$$

This proves our theorem. $\qquad\qquad\qquad\qquad\qquad\qquad\qquad\qquad\qquad$ □

Corollary 6.8. *Suppose $f(\cdot) \in C([-\pi, \pi]; \mathbb{C})$ satisfying the following:*

$$\langle f, e_n \rangle = 0, \qquad \forall n \in \mathbb{Z}.$$

Then $f(\cdot) = 0$.

Proof. We can find a sequence $T_m(\cdot)$ of trigonometric polynomials of the following form:

$$T_m(x) = \sum_{n=-N_m}^{N_m} c_{mn} e_n(x),$$

such that

$$\|f(\cdot) - T_m(\cdot)\|_\infty < \frac{1}{m}, \qquad \forall m \geqslant 1.$$

Then,

$$\|f(\cdot)\|_2^2 = \langle f, f - T_m \rangle = \int_{-\pi}^{\pi} |f(x)|\,|f(x) - T_m(x)|dx$$
$$\leqslant 2\pi \|f\|_\infty \|f - T_m\|_\infty < \frac{2\pi}{m}\|f\|_\infty \to 0.$$

Hence, one must have $f(\cdot) = 0$. $\qquad\square$

The above result tells us that if $f(\cdot) \in C([-\pi, \pi]; \mathbb{R})$ is perpendicular to every $e_n(\cdot)$, then it is necessary that $f(\cdot) = 0$. This means that the orthonormal system $\{e_n\}$ is *complete*, in some sense.

6.3 *Fourier series*

Now, if $f(\cdot)$ is given by (6.3), then it is continuous and periodic with period 2π. We observe the following:

$$\int_{-\pi}^{\pi} f(x)e^{-imx}dx = \sum_{n=-N}^{N} c_n \int_{-\pi}^{\pi} e^{i(n-m)x}dx = 2\pi c_m.$$

Hence,

$$c_n = \frac{1}{2\pi}\int_{-\pi}^{\pi} f(x)e^{-inx}dx = \frac{1}{\sqrt{2\pi}}\langle f, e_n \rangle, \quad -N \leqslant n \leqslant N. \qquad (6.7)$$

In other words, such an $f(\cdot)$ admits the following representation:

$$f(x) = \frac{1}{\sqrt{2\pi}} \sum_{n=-N}^{N} \langle f, e_n \rangle e_n.$$

Now, for any $f : [-\pi, \pi] \to \mathbb{C}$ with $|f(\cdot)| \in L_R([-\pi, \pi]; \mathbb{R})$, we define

$$\widehat{f}(n) = \langle f, e_n \rangle = \frac{1}{\sqrt{2\pi}}\int_{-\pi}^{\pi} f(x)e^{-inx}dx, \qquad n \in \mathbb{Z}, \qquad (6.8)$$

which is called the n-th *Fourier coefficient* of $f(\cdot)$. We write

$$f(\cdot) \sim \sum_{n=-\infty}^{\infty} \widehat{f}(n)e_n = \frac{1}{\sqrt{2\pi}} \sum_{n=-\infty}^{\infty} \widehat{f}(n)e^{inx},$$

indicating the relation between $f(\cdot)$ and the coefficients of the trigonometric series. The right-hand side of the above is called the Fourier series of $f(\cdot)$. Note that $f(\cdot)$ is not necessarily bounded, not necessarily continuous. Hence, we usually might not have

$$f(\cdot) = \sum_{n=-\infty}^{\infty} \widehat{f}(n)e_n. \tag{6.9}$$

However, we have the following result.

Theorem 6.9. (Fourier) *Let* $f(\cdot) \in C([-\pi, \pi]; \mathbb{C})$. *Then its Fourier series* $\sum\limits_{n=-\infty}^{\infty} \widehat{f}(n)e_n$ *converges in* L^2-*norm to* $f(\cdot)$, *i.e.*,

$$\lim_{N \to \infty} \left\| f - \sum_{n=-N}^{N} \widehat{f}(n)e_n \right\|_2^2$$
$$\equiv \lim_{N \to \infty} \int_{-\pi}^{\pi} \left| f(x) - \sum_{n=-N}^{N} \widehat{f}(n)e_n(x) \right|^2 dx = 0. \tag{6.10}$$

Further, if

$$\sum_{n=-\infty}^{\infty} |\widehat{f}(n)| < \infty, \tag{6.11}$$

then the Fourier series $\sum\limits_{n=-\infty}^{\infty} \widehat{f}(n)e_n$ *converges to* $f(\cdot)$ *uniformly, i.e.*,

$$\lim_{N \to \infty} \left\| f - \sum_{n=-N}^{N} \widehat{f}(n)e_n \right\|_\infty = 0.$$

Proof. For any $\varepsilon > 0$, by Weierstrass Theorem, there exists a trigonometric polynomial

$$T(\cdot) = \sum_{n=-N_0}^{N_0} c_n e_n,$$

such that

$$\|f - T\|_\infty \leqslant \varepsilon.$$

Now, for any $N > N_0$, let

$$F_N(x) = \sum_{n=-N}^{N} \widehat{f}(n) e_n(x).$$

Then for any $|m| \leqslant N$, we have

$$\langle f - F_N, e_m \rangle = \langle f, e_m \rangle - \sum_{n=-N}^{N} \widehat{f}(n) \langle e_n, e_m \rangle = \widehat{f}(m) - \widehat{f}(m) = 0.$$

This implies that

$$\langle f - F_N, F_N - T \rangle = \langle f - F_N, \sum_{n=-N_0}^{N_0} (\widehat{f}(n) - c_n) e_n \rangle$$

$$+ \langle f - F_N, \Big(\sum_{n=-N}^{-(N_0+1)} + \sum_{n=N_0+1}^{N} \Big) \widehat{f}(n) e_n \rangle = 0.$$

Consequently,

$$\|f - T\|_2^2 = \|f - F_N + F_N - T\|_2^2$$
$$= \|f - F_N\|_2^2 + \|F_N - T\|_2^2 + 2\langle f - F_N, F_N - T \rangle$$
$$= \|f - F_N\|_2^2 + \|F_N - T\|_2^2.$$

Hence,

$$\|f - F_N\|_2^2 \leqslant \|f - T\|_2^2 \leqslant 2\pi\varepsilon.$$

This proves

$$\lim_{N \to \infty} \|f - F_N\|_2 = 0.$$

Now, suppose (6.11) holds, then by Weierstrass M-test, we see that the Fourier series is convergent uniformly to some function $F(\cdot)$, which implies that

$$\Big\| \sum_{n=-N}^{N} \widehat{f}(n) e_n - F \Big\|_2^2 \leqslant 4\pi^2 \Big\| \sum_{n=-N}^{N} \widehat{f}(n) e_n - F \Big\|_\infty^2 \to 0,$$

uniformly, as $N \to \infty$. By the uniqueness of the L^2-limit, we must have $F(\cdot) = f(\cdot)$, proving our theorem. $\qquad\square$

Because of the above result, for any $f(\cdot) \in C([-\pi, \pi]; \mathbb{C})$, we may write

$$f(x) = \sum_{n=-\infty}^{\infty} \widehat{f}(n) e_n(x) \equiv \frac{1}{\sqrt{2\pi}} \sum_{n=-\infty}^{\infty} \widehat{f}(n) e^{inx},$$

which is understood as (6.10). Further, if $f(\cdot) \in C([-\pi, \pi]; \mathbb{R})$ and the Fourier series is uniformly convergent on $[a, b] \subseteq [-\pi, \pi]$, then the above equality holds for $x \in [a, b]$. Moreover, we have the following result.

Theorem 6.10. (Plancherel) *For any* $f(\cdot) \in C([-\pi, \pi]; \mathbb{C})$, *the series* $\sum_{n=-\infty}^{\infty} |\widehat{f}(n)|^2$ *is convergent and the following Parseval's equality holds:*

$$\|f\|_2^2 = \sum_{n=-\infty}^{\infty} |\widehat{f}(n)|^2. \tag{6.12}$$

Proof. For any $\varepsilon > 0$, by Fourier Theorem, there exists an $N_0 \geqslant 1$ such that

$$\left\| f - \sum_{n=-N}^{N} \widehat{f}(n) e_n \right\|_2 \leqslant \varepsilon, \qquad \forall N \geqslant N_0.$$

This leads to

$$\|f\|_2 - \varepsilon \leqslant \left\| \sum_{n=-N}^{N} \widehat{f}(n) e_n \right\|_2 \leqslant \|f\|_2 + \varepsilon.$$

On the other hand, since $\{e_n\}$ is an orthonormal system, it is clear that

$$\left\| \sum_{n=-N}^{N} \widehat{f}(n) e_n \right\|_2^2 = \sum_{n=-N}^{N} |\widehat{f}(n)|^2.$$

Thus,

$$(\|f\|_2 - \varepsilon)^2 \leqslant \sum_{n=-N}^{N} |\widehat{f}(n)|^2 \leqslant (\|f\|_2 + \varepsilon)^2.$$

Consequently,

$$(\|f\|_2 - \varepsilon)^2 \leqslant \varliminf_{N \to \infty} \sum_{n=-N}^{N} |\widehat{f}(n)|^2 \leqslant (\|f\|_2 + \varepsilon)^2.$$

Since $\varepsilon > 0$ is arbitrary, we obtain the Parseval's equality. $\qquad\square$

Corollary 6.11. *Let* $f : [-\pi, \pi] \to \mathbb{R}$ *be Riemann integrable with* $|f(\cdot)|^2$ *also being Riemann integrable. Then (6.10) holds.*

Proof. We can find a sequence of continuous functions $f^m(\cdot)$ such that

$$\lim_{m \to \infty} \|f(\cdot) - f^m(\cdot)\|_2 = 0.$$

From the above theorem, we have

$$\sum_{n=-\infty}^{\infty} |\widehat{f^m}(n) - \widehat{f^\ell}(n)|^2 = \|f^m(\cdot) - f^\ell(\cdot)\|_2^2 \to 0, \qquad m, \ell \to \infty.$$

Note that

$$|\widehat{f^m}(n) - \widehat{f}(n)| \leqslant \frac{1}{\sqrt{2\pi}} \int_{-\pi}^{\pi} |f^m(x) - f(x)| dx$$

$$\leqslant \frac{1}{2\pi} \Big(\int_{-\pi}^{\pi} |f^m(x) - f(x)|^2 dx \Big)^{\frac{1}{2}} \to 0.$$

By Fatou's Lemma (Theorem 3.6, (iii)), we obtain

$$\sum_{n=-\infty}^{\infty} |\widehat{f^m}(n) - \widehat{f}(n)|^2 \leqslant \|f^m(\cdot) - f(\cdot)\|_2^2.$$

Hence,

$$\|F_N^m(\cdot) - F_N(\cdot)\|_2^2 \equiv \frac{1}{2\pi} \int_{-\pi}^{\pi} \Big| \sum_{n=-N}^{N} [\widehat{f^m}(n) - \widehat{f}(n)] e^{-inx} \Big|^2 dx$$

$$= \sum_{n=-N}^{N} |\widehat{f^m}(n) - \widehat{f}(n)|^2 \leqslant \sum_{n=-\infty}^{\infty} |\widehat{f^m}(n) - \widehat{f}(n)|^2 \leqslant \|f^m(\cdot) - f(\cdot)\|_2^2.$$

Also,

$$\sum_{n=-\infty}^{\infty} |\widehat{f}(n)|^2 = \sum_{n=-\infty}^{\infty} \Big(2|\widehat{f^m}(n)|^2 + 2|\widehat{f^m}(n) - \widehat{f}(n)|^2 \Big)$$

$$= 2\|f^m(\cdot)\|_2^2 + 2\|f^m(\cdot) - f(\cdot)\|_2^2 \leqslant 4\|f(\cdot)\|_2^2 + 6\|f^m(\cdot) - f(\cdot)\|_2^2 < \infty.$$

Then

$$\|f(\cdot) - F_N(\cdot)\|_2^2$$

$$\leqslant 3\Big(\|f(\cdot) - f^m(\cdot)\|_2^2 + \|f^m(\cdot) - F_N^m(\cdot)\|_2^2 + \|F_N^m(\cdot) - F_N(\cdot)\|_2^2 \Big)$$

$$\leqslant 6\|f(\cdot) - f^m(\cdot)\|_2^2 + \|f^m(\cdot) - F_N^m(\cdot)\|_2^2$$

Consequently,

$$\overline{\lim_{N \to \infty}} \|f(\cdot) - F_N(\cdot)\|_2^2 \leqslant 6\|f(\cdot) - f^m(\cdot)\|_2^2.$$

Therefore, one must have (6.10). $\qquad \square$

The above result tells us that when $f(\cdot)$, together with $|f(\cdot)|$ being Riemann integrable, we can write (6.9), as long as it is understood in the sense of (6.10).

From the relation (6.2), we see that for any $f(\cdot) \in C([-\pi, \pi]; \mathbb{R})$, one may write (see (6.8))

$$f(x) = \frac{1}{\sqrt{2\pi}} \sum_{n=-\infty}^{\infty} \widehat{f}(n) e^{inx} = \frac{1}{2\pi} \sum_{n=-\infty}^{\infty} \left(\int_{-\pi}^{\pi} f(y) e^{-iny} dy \right) e^{inx}.$$

Note that

$$\left(\int_{-\pi}^{\pi} f(y) e^{-iny} dy \right) e^{inx} + \left(\int_{-\pi}^{\pi} f(y) e^{iny} dy \right) e^{-inx}$$

$$= \left(\int_{-\pi}^{\pi} f(y)(e^{-iny} + e^{iny}) dy \right) \cos nx + \left(\int_{-\pi}^{\pi} f(y) i(e^{-iny} - e^{iny}) dy \right) \sin nx$$

$$= 2 \left(\int_{-\pi}^{\pi} f(y) \cos nydy \right) \cos nx + 2 \left(\int_{-\pi}^{\pi} f(y) \sin nydy \right) \sin nx.$$

Hence,

$$f(x) = \frac{1}{2\pi} \left(\int_{-\pi}^{\pi} f(y) dy \right) + \frac{1}{\pi} \sum_{n=1}^{\infty} \left(\int_{-\pi}^{\pi} f(y) \cos nydy \right) \cos nx$$

$$+ \frac{1}{\pi} \sum_{n=1}^{\infty} \left(\int_{-\pi}^{\pi} f(y) \sin nydy \right) \sin nx$$

$$\equiv a_0 + \sum_{n=1}^{\infty} a_n \cos nx + \sum_{n=1}^{\infty} b_n \sin nx,$$

with

$$\begin{cases} a_0 = \dfrac{1}{2\pi} \displaystyle\int_{-\pi}^{\pi} f(y) dy, \qquad a_n = \dfrac{1}{\pi} \displaystyle\int_{-pi}^{\pi} f(y) \cos nydy, \\[4mm] b_n = \dfrac{1}{\pi} \displaystyle\int_{-\pi}^{\pi} f(y) \sin nydy, \qquad n \geqslant 1. \end{cases} \tag{6.13}$$

From the above, we can restate Theorems 6.9 and 6.10 as follows.

Theorem 6.12. *Let* $f : [-\pi, \pi] \to \mathbb{R}$, *together with* $|f(\cdot)|^2$ *be Riemann integrable, let* a_n, b_n *be defined by* (6.13). *Then the following are true:*

(i) *The Parseval's equality holds:*

$$2\pi a_0^2 + \pi \sum_{n=1}^{\infty} \left(a_n^2 + b_n^2 \right) = \|f\|_2^2. \tag{6.14}$$

(ii) *The Fourier series* $a_0 + \sum_{n=1}^{\infty}(a_n \cos nx + b_n \sin nx)$ *converges to* $f(\cdot)$ *in* L^2-*norm, i.e.,*

$$\lim_{N\to\infty}\left\|\left(a_0 + \sum_{n=1}^{N}(a_n \cos n\cdot + b_n \sin n\cdot)\right) - f(\cdot)\right\|_2 = 0.$$

(iii) *If* $\sum_{n=1}^{\infty}\left(|a_n| + |b_n|\right) < \infty$ *and* $f(\cdot) \in C([-\pi, \pi]; \mathbb{R})$, *then*

$$\lim_{N\to\infty}\left\|\left(a_0 + \sum_{n=1}^{N}(a_n \cos n\cdot + b_n \sin n\cdot)\right) - f(\cdot)\right\|_\infty = 0,$$

i.e., the Fourier series converges to $f(\cdot)$ *uniformly.*

It is interesting that in order for a trigonometric series to be a Fourier series of some continuous function, (6.14) has to be true, in particular, the series on the left-hand side must be convergent.

Exercises

1. Let $f : [-\pi, \pi] \to \mathbb{R}$ be continuous. Let

$$a_0 = \frac{1}{2\pi}\int_{-\pi}^{\pi} f(x)dx, \quad a_k = \frac{1}{\pi}\int_{-\pi}^{\pi} f(x)\cos kx\,dx,$$

$$b_k = \frac{1}{\pi}\int_{-\pi}^{\pi} f(x)\sin kx\,dx, \quad k \geqslant 1.$$

Let $-\pi < c < x < \pi$. Show that

$$\int_c^x f(\xi)d\xi = a_0(x - c) + \sum_{k=1}^{\infty}\int_c^x \left(a_k \cos k\xi + b_k \sin k\xi\right)d\xi.$$

2. Show that the series $\sum_{k=1}^{\infty}\dfrac{b_k}{k}$ is convergent, where b_k is defined as in the above problem for some continuous function $f(\cdot)$ on $[-\pi, \pi]$.

3. State and prove the results of Fourier series for $C([a, b]; \mathbb{R})$, i.e., replace $[-\pi, \pi]$ by general $[a, b]$.

4. Let $f : [-\pi, \pi] \to \mathbb{R}$ be continuously differentiable with $f(-\pi) = f(\pi)$. Suppose

$$f(x) \sim a_0 + \sum_{k=1}^{\infty}\left(a_k \cos kx + b_k \sin kx\right).$$

Show that

$$f'(x) \sim \sum_{k=1}^{\infty} \frac{d}{dx}\Big(a_k \cos kx + b_k \sin kx\Big).$$

5*. Show that $\displaystyle\sum_{k=2}^{\infty} \frac{\sin kx}{\ln k}$ is convergent on $[0, 2\pi]$, and is convergent uniformly on any $[\delta, 2\pi - \delta]$ for any small $\delta > 0$. Is this a Fourier series of some continuous function on $(0, 2\pi)$? Why?

Appendix

The Real Number System

1 Natural Numbers

1.1 *Peano axioms*

Let us begin with some basic notions.

- A collection of certain type objects is called a *set*.[1] If A is a set, and an object a is in A, we denote $a \in A$. In this case, we call a an *element* in A. If a is not an element of A, we denote $a \notin A$. If A contains at least one element, we say that A is *non-empty*. If A contains no element, we say that A is *empty*, denoted by $A = \varnothing$.

- Let A and B be two sets. Denote $A \setminus B$, called the *difference* of A and B, to be the set that contains all the elements in A but not in B. If any element of A is in B, we say that A is a *subset* of B, denoted by $A \subseteq B$, or equivalently, $B \supseteq A$. In the case that A is a subset of B and B is also a subset of A, we say that A and B are *equal*, denoted by $A = B$. If A and B are different, by which we mean that at least one element in A is not in B, or at least one element in B is not in A, then we denote $A \neq B$. In the case that $A \subseteq B$ and $A \neq B$, we call A a *proper subset* of B, denoted by $A \subset B$.

- Let A and B be two sets. Let $f(\cdot)$ be a rule that assigns each element $a \in A$ a unique element $b \in B$, denoted by $f(a) = b$. We call such an f a *map* from A to B, denoted by $f : A \to B$.

Now, let $\overline{\mathbb{N}}$ be a non-empty set. On $\overline{\mathbb{N}}$, we introduce a *binary relation* called *equal*, denoted by "$=$", which has the following properties:

---

[1] We are satisfied with the definition of set at such a level and we have no intention to introduce axioms for that.

- (Reflexivity) For all $n \in \overline{\mathbb{N}}$, $n = n$;

- (Symmetry) For all $n, m \in \overline{\mathbb{N}}$, if $n = m$, then $m = n$;

- (Transitivity) For all $n, m, \ell \in \overline{\mathbb{N}}$, if $n = m$ and $m = \ell$, then $n = \ell$.

We now introduce the *Peano Axioms* for the set $\overline{\mathbb{N}}$.

Axiom 1. A map $\sigma : \overline{\mathbb{N}} \to \overline{\mathbb{N}}$ is defined such that for all $n, m \in \overline{\mathbb{N}}$, $n = m$ if and only if $\sigma(n) = \sigma(m)$. The map σ is called a *successor function*, and for every $n \in \overline{\mathbb{N}}$, $\sigma(n) \in \overline{\mathbb{N}}$ is called the *successor* of n.

Axiom 2. There is a special element in $\overline{\mathbb{N}}$, denoted by 0, such that for all $n \in \overline{\mathbb{N}}$, $\sigma(n) \neq 0$.

Axiom 3. (Principle of Mathematical Induction) Let K be a non-empty set having two properties: (i) $0 \in K$, and (ii) $n \in K$ implies $\sigma(n) \in K$. Then $K \supseteq \overline{\mathbb{N}}$.

Let $\overline{\mathbb{N}}$ satisfy the above Peano Axioms, and let $\mathbb{N} = \overline{\mathbb{N}} \setminus \{0\}$. We call $\mathbb{N}$ the *natural number set* and any element in $\mathbb{N}$ is called a *natural number*.

Another way of stating Axiom 3 is as follows.

Axiom 3′. If $P(n)$ is a statement depending on $n \in \overline{\mathbb{N}}$ such that

- $P(0)$ is true, and

- for every $n \in \overline{\mathbb{N}}$, if $P(n)$ is true, so is $P(\sigma(n))$,

then $P(n)$ is true for every $n \in \overline{\mathbb{N}}$.

We will denote

$$\sigma(0) = 1, \quad \sigma(1) = 2, \quad \sigma(2) = 3, \quad \sigma(3) = 4, \cdots,$$

and so on. By Axioms 1 and 2, we see that

$$\mathbb{K} \equiv \{0, 1, 2, \cdots\} = \{0, \sigma(0), \sigma(\sigma(0)), \cdots\} \subseteq \overline{\mathbb{N}}.$$

On the other hand, let $P(n)$ be the statement "$n \in \mathbb{K}$". Then $P(0)$ is true and if $P(n)$ is true, $P(\sigma(n))$ is true. Hence, by Axiom 3′, $P(n)$ is true for all $n \in \overline{\mathbb{N}}$, which means $\overline{\mathbb{N}} \subseteq \mathbb{K}$. Therefore, we obtain $\overline{\mathbb{N}} = \mathbb{K}$, i.e.,

$$\overline{\mathbb{N}} = \{0, 1, 2, \cdots\}, \qquad \mathbb{N} = \{1, 2, 3, \cdots\}. \tag{1.1}$$

1.2 Addition and multiplication

In this subsection, we introduce two *binary operations* in $\overline{\mathbb{N}}$. First, let us introduce the following.

Definition 1.1. For any given $m \in \overline{\mathbb{N}}$, define

$$0 + m = m. \tag{1.2}$$

Inductively, suppose $n + m$ is defined. Then we define

$$\sigma(n) + m = \sigma(n + m). \tag{1.3}$$

The above defined operation "$+$" is called the *addition* on $\overline{\mathbb{N}}$.

For the addition defined above, we have the following result.

Lemma 1.2. (i) *For any $n \in \overline{\mathbb{N}}$,*

$$n + 0 = n. \tag{1.4}$$

(ii) *For any $n, m \in \overline{\mathbb{N}}$,*

$$n + \sigma(m) = \sigma(n + m). \tag{1.5}$$

Proof. (i) Use induction on n.

For $n = 0$: By (1.2), we have

$$0 + 0 = 0.$$

Now, suppose $n + 0 = n$. By (1.3) and induction hypothesis,

$$\sigma(n) + 0 = \sigma(n + 0) = \sigma(n),$$

proving (1.4).

(ii) Fix m, using induction on n.

For $n = 0$: By (1.2), one has

$$\sigma(0 + m) = \sigma(m) = 0 + \sigma(m).$$

Hence, the statement holds for $n = 0$. Now suppose

$$n + \sigma(m) = \sigma(n + m).$$

Then the above, together with (1.3), gives

$$\sigma(n) + \sigma(m) = \sigma\big(n + \sigma(m)\big) = \sigma\big(\sigma(n + m)\big) = \sigma(\sigma(n) + m).$$

This completes the induction. $\qquad\square$

Combining the above lemma with the definition of addition, we have

$$0 + n = n + 0 = n, \qquad \forall n \in \mathbb{N}, \tag{1.6}$$

and

$$\sigma(n) + m = n + \sigma(m) = \sigma(n + m), \quad \forall n, m \in \mathbb{N}. \tag{1.7}$$

The above leads to the following result on the addition of the elements in $\overline{\mathbb{N}}$.

Proposition 1.3. *For any* $n, m, \ell \in \mathbb{N}$,

(i) (Commutativity) $n + m = m + n$.

(ii) (Associativity) $(n + m) + \ell = n + (m + \ell)$.

(iii) (Cancellation Property) $n + m = n + \ell$ *if and only if* $m = \ell$.

Proof. (i) Fix an $m \in \mathbb{N}$, we use induction on n.

For $n = 0$: It follows from (1.6).

Now, suppose

$$n + m = m + n, \qquad \forall m \in \mathbb{N}.$$

Then it follows from (1.7) that

$$\sigma(n) + m = \sigma(n + m) = \sigma(m + n) = m + \sigma(n).$$

This proves (i).

(ii) Fix $m, \ell \in \mathbb{N}$, we use induction on n.

For $n = 0$: By the definition of addition, we have

$$0 + (m + \ell) = m + \ell = (0 + m) + \ell.$$

Now, suppose

$$(n + m) + \ell = n + (m + \ell).$$

By the definition of addition, and induction hypothesis, we have

$$\big(\sigma(n) + m\big) + \ell = \sigma(n + m) + \ell = \sigma\big((n + m) + \ell\big)$$
$$= \sigma\big(n + (m + \ell)\big) = \sigma(n) + (m + \ell).$$

Hence, (ii) follows.

(iii) Use induction on n.

For $n = 0$: By definition of addition, one clearly has

$$0 + m = 0 + \ell \quad \Longleftrightarrow \quad m = \ell.$$

Thus, the statement holds for $n = 0$. Now, suppose that

$$n + m = n + \ell \quad \Longleftrightarrow \quad m = \ell. \tag{1.8}$$

We want to show that

$$\sigma(n) + m = \sigma(n) + \ell \quad \Longleftrightarrow \quad m = \ell. \tag{1.9}$$

By (1.3), the left-hand side of (1.9) implies

$$\sigma(n + m) = \sigma(n) + m = \sigma(n) + \ell = \sigma(n + \ell).$$

By Axiom 2, one must have the left-hand side of (1.8) which leads to $m = \ell$ by induction hypothesis (1.8). Conversely, if $m = \ell$, then the left-hand side of (1.8) holds (by induction hypothesis). Hence, by Axiom 1,

$$\sigma(m) = \sigma(\ell), \qquad \sigma(n + m) = \sigma(n + \ell).$$

Consequently, by (1.3),

$$\sigma(n) + m = \sigma(n + m) = \sigma(n + \ell) = \sigma(n) + \ell,$$

proving our claim. $\qquad\qquad\square$

From Lemma 1.2, we also have the following interesting corollary.

Corollary 1.4. *For any* $n \in \overline{\mathbb{N}}$,

$$\sigma(n) = n + 1. \tag{1.10}$$

Proof. By (1.4)–(1.5), we have

$$\sigma(n) = \sigma(n + 0) = n + \sigma(0) = n + 1.$$

This proves our conclusion. $\qquad\qquad\square$

Because of the above corollary, from now on, we will use $n + 1$ instead of $\sigma(n)$. We now introduce another binary operation on $\overline{\mathbb{N}}$.

Definition 1.5. Let $m \in \overline{\mathbb{N}}$. Define

$$0 \times m = 0, \tag{1.11}$$

Suppose $n \times m$ is defined. Then define

$$(n + 1) \times m = (n \times m) + m. \tag{1.12}$$

The above defined operation "$\times$" is called the *multiplication* in $\overline{\mathbb{N}}$. Similar to Lemma 1.2, we have the following lemma.

Lemma 1.6. (i) *For any $n \in \overline{\mathbb{N}}$,*

$$n \times 0 = 0, \qquad n \times 1 = n. \tag{1.13}$$

(ii) *For any $m, n \in \overline{\mathbb{N}}$,*

$$n \times (m + 1) = (n \times m) + n. \tag{1.14}$$

Proof. (i) We prove the first equality in (1.13) by induction.

For $n = 0$: By definition, $0 \times 0 = 0$.

Suppose $n \times 0 = 0$. Then by definition and induction hypothesis,

$$(n + 1) \times 0 = (n \times 0) + 0 = 0 + 0 = 0.$$

This proves the first equality in (1.13).

Now, we prove the second equality in (1.13) by induction.

For $n = 0$: By (1.11), we have

$$0 \times 1 = 0.$$

Suppose the conclusion holds for n. Then by definition and induction hypothesis, we have

$$(n + 1) \times 1 = (n \times 1) + 1 = n + 1,$$

proving the second equality in (1.13).

(ii) Fix m, use induction on n.

For $n = 0$: We need to show that

$$0 \times (m + 1) = (0 \times m) + 0,$$

which is true since both sides are 0.

Now suppose

$$n \times (m + 1) = (n \times m) + n.$$

Then for $n + 1$, by definition of multiplication, the induction hypothesis, and Proposition 1.3,

$$(n + 1) \times (m + 1) = n \times (m + 1) + (m + 1) = (n \times m) + n + (m + 1)$$
$$= (n \times m) + m + (n + 1) = (n + 1) \times m + (n + 1).$$

Hence, our conclusion follows. $\qquad\square$

From the definition of multiplication and the above proposition, we have

$$0 \times m = m \times 0 = 0, \quad 1 \times m = m \times 1 = m, \qquad \forall m \in \overline{\mathbb{N}}. \tag{1.15}$$

Now, we are able to prove the following result.

Proposition 1.7. *For any* $n, m, \ell \in \overline{\mathbb{N}}$,

(i) (Commutativity) $n \times m = m \times n$.

(ii) (Distributivity) $n \times (m + \ell) = n \times m + n \times \ell$.

(iii) (Associativity) $(n \times m) \times \ell = n \times (m \times \ell)$.

Proof. (i) Fix m, and use induction on n.

For $n = 0$: It is trivial since both sides are zero (see (1.15)).

Suppose

$$n \times m = m \times n.$$

Then by (1.12) and (1.14), one has

$$(n + 1) \times m = (n \times m) + m = (m \times n) + m = m \times (n + 1).$$

This proves (i).

(ii) Fix $n, m \in \overline{\mathbb{N}}$ and use induction on ℓ.

For $\ell = 0$: We need to show that

$$n \times (m + 0) = n \times m + n \times 0.$$

This is true since both sides are equal to $n \times m$. Suppose

$$n \times (m + \ell) = n \times m + n \times \ell.$$

Then by (1.14),

$$n \times \big(m + (\ell + 1)\big) = n \times \big((m + \ell) + 1\big) = n \times (m + \ell) + n$$
$$= n \times m + n \times \ell + n = n \times m + n \times (\ell + 1).$$

This proves (ii).

(iii) Fix $m, \ell \in \overline{\mathbb{N}}$ and use induction on n.

For $n = 0$: We need to show

$$(0 \times m) \times \ell = 0 \times (m \times \ell),$$

which is true since both sides are zero.

Now suppose

$$(n \times m) \times \ell = n \times (m \times \ell).$$

Then by the proved (ii),

$$\big((n+1) \times m\big) \times \ell = \big((n \times m) + m\big) \times \ell = (n \times m) \times \ell + m \times \ell$$
$$= n \times (m \times \ell) + m \times \ell = (n+1) \times (m \times \ell).$$

This completes the proof. □

1.3   *Orders*

We now look at another important feature of elements in $\bar{\mathbb{N}}$, which is described by the following binary relation.

Definition 1.8. Let $n, m \in \bar{\mathbb{N}}$. We say that $n \geqslant m$ (or $m \leqslant n$) if there exists an $\ell \in \bar{\mathbb{N}}$ such that

$$n = m + \ell,$$

and we say that $n > m$ (or $m < n$) if

$$n \geqslant m, \qquad n \neq m.$$

In particular, when $n > 0$, we say that n is *positive*.

Let us now present the following simple lemma.

Lemma 1.9. (i) *Let $n \in \bar{\mathbb{N}}$. Then, $n \geqslant 0$; and $n > 0$ if and only if $n \neq 0$.*

(ii) *For any $n \in \mathbb{N}$, there exists a unique $m \in \bar{\mathbb{N}}$ such that $n = m + 1$.*

(iii) *Let $n, m \in \bar{\mathbb{N}}$. Then $n > m$ if and only if $n = m + \ell$, for some $\ell > 0$; and if and only if $n \geqslant m + 1$.*

(iv) *Let $n, m \in \bar{\mathbb{N}}$. Then $n \geqslant m$ if and only if $n + 1 > m$.*

(v) *Let $n, m \in \bar{\mathbb{N}}$. Then $n + m > 0$ if and only if either $n > 0$ or $m > 0$.*

Proof. (i) For any $n \in \bar{\mathbb{N}}$, since $n = 0 + n$, by definition, $n \geqslant 0$; Further, if $n \neq 0$, then by definition, $n > 0$. Conversely, if $n > 0$, by definition $n \neq 0$.

(ii) By Axiom 1, if $m \in \bar{\mathbb{N}}$ exists such that $n = m + 1$, it must be unique. We use induction on n.

For $n = 1$: We may take $m = 0$ since $1 = 0 + 1$.

Now, suppose $n > 0$ and there exists a unique $m \in \overline{\mathbb{N}}$ such that $n = m + 1$. Then, for $n + 1$, it suffices to take $\ell = m + 1$ to get

$$n + 1 = (m + 1) + 1 = \ell + 1.$$

This completes the induction.

(iii) By definition, $n > m$ if and only if $n = m + \ell$ for some $\ell \in \overline{\mathbb{N}}$ and $n \neq m$. Hence, $\ell \neq 0$, or $\ell > 0$. (Otherwise, $n = m$.)

Next, suppose $n = m + \ell$ with $\ell > 0$. Since $\ell > 0$, by (ii), one has $\ell = \bar{\ell} + 1$ for some $\bar{\ell} \in \overline{\mathbb{N}}$. Then

$$n = m + \ell = m + (\bar{\ell} + 1) = (m + 1) + \bar{\ell},$$

which means $n \geqslant m + 1$.

Conversely, if $n \geqslant m + 1$, then there exists an $\ell \in \overline{\mathbb{N}}$ such that

$$n = (m + 1) + \ell = m + (\ell + 1),$$

and $(\ell + 1)$ is positive, which gives $n > m$.

(iv) The proof is similar to that of (iii), and it is left to the readers.

(v) Suppose $n + m > 0$, but $n = m = 0$. Then $n + m = 0 + 0 = 0$, a contradiction. Conversely, if, say, $n > 0$, then we show $n + m > 0$ by induction on m.

For $m = 0$: In this case,

$$n + m = n + 0 = 0 + n = n > 0.$$

Now, suppose that $n + m > 0$. We want to show that $n + (m + 1) > 0$. In fact, by the associativity of addition, one has

$$n + (m + 1) = (n + m) + 1,$$

which cannot be 0 (by Axiom 2). Hence, $n + (m + 1) > 0$. This completes the induction. □

The following proposition collects the three most important properties of the relation "$\geqslant$" introduced in Definition 1.8.

Proposition 1.10. *Let $n, m, \ell \in \overline{\mathbb{N}}$. Then*

(i) (Reflexivity) $n \geqslant n$.

(ii) (Transitivity) *If $n \geqslant m$, $m \geqslant \ell$, then $n \geqslant \ell$.*

(iii) (Anti-symmetry) *If $n \geqslant m$ and $m \geqslant n$, then $n = m$.*

Proof. (i) Since $n = n + 0$ with $0 \in \overline{\mathbb{N}}$, we have $n \geqslant n$.

(ii) Under our conditions,

$$n = m + \bar{m}, \qquad m = \ell + \bar{\ell},$$

for some $\bar{m}, \bar{\ell} \in \overline{\mathbb{N}}$. Then

$$n = m + \bar{m} = \ell + \bar{\ell} + \bar{m}.$$

Since $\bar{m} + \bar{\ell} \in \overline{\mathbb{N}}$, we see that $n \geqslant \ell$, by definition.

(iii) If $n \neq m$, then together with $n \geqslant m$, we have $n > m$ and thus there exists an $\bar{m} \in \mathbb{N}$ such that

$$n = m + \bar{m}.$$

Likewise, there exists an $\bar{n} \in \mathbb{N}$ such that

$$m = n + \bar{n} = m + \bar{m} + \bar{n}.$$

By the cancellation property of addition (see Proposition 1.3, (iii)), we obtain

$$\bar{n} + \bar{m} = 0,$$

with both $\bar{n} > 0$ and $\bar{m} > 0$. This contradicts Lemma 1.9, (v).   □

Proposition 1.11. (Trichotomy) *Let $n, m \in \overline{\mathbb{N}}$. Then exactly one of the following holds:*

$$n > m, \quad n = m, \quad n < m. \tag{1.16}$$

Proof. First, we show that for any $n, m \in \overline{\mathbb{N}}$, at most one of relation in (1.16) can hold. In fact, if $n > m$, then both $n = m$ and $n < m$ fail; if $n < m$, then both $n = m$ and $n > m$ fail; and if $n = m$, then both $n > m$ and $n < m$ fail.

Next, we show that at least one of the three is true. To show that, we fix m and use induction on n.

For $n = 0$: For any $m \in \overline{\mathbb{N}}$, $m \geqslant 0$ since $m = m + 0$ with $0 \in \overline{\mathbb{N}}$. So either $m = 0$, or $m > 0$, i.e., either $m = n$ or $m > n$, with $n = 0$.

Now, suppose the statement holds for $n \in \overline{\mathbb{N}}$, i.e., one of the three relations holds between n and m.

In the case $n > m$, we have $(n + 1) > n > m$, leading to $(n + 1) > m$.

In the case $n = m$, we have $(n + 1) > n = m$ leading to $(n + 1) > m$.

In the case $n < m$, by (iii) of Proposition 1.9, we have $(n + 1) \leqslant m$, which means $m = (n + 1) + \ell$ for some $\ell \in \bar{\mathbb{N}}$. If $\ell = 0$, then $m = (n + 1)$. If $\ell > 0$, then

$$(n + 1) < m.$$

Either case leads to a completion of the proof. $\qquad\square$

The following proposition collects some compatibility properties of the addition and multiplication with the order and the equality.

Proposition 1.12. *Let $n, m, \ell \in \bar{\mathbb{N}}$.*

(i) *(Order Preservation under Addition) $n \geqslant m$ if and only if $n + \ell \geqslant m + \ell$.*

(ii) *(Order Preservation under Multiplication) If $\ell > 0$, then $n > m$ if and only if $n \times \ell > m \times \ell$. In particular (taking $m = 0$), $n \times \ell > 0$ if and only if $n, \ell > 0$.*

(iii) *(Cancellation Property) If $\ell > 0$, then $n \times \ell = m \times \ell$ if and only if $n = m$.*

Proof. (i) By definition, $n \geqslant m$ if and only if

$$n = m + \bar{m}, \qquad\qquad (1.17)$$

for some $\bar{m} \in \bar{\mathbb{N}}$. Then, by the associativity and commutativity of addition,

$$n + \ell = (m + \bar{m}) + \ell = (m + \ell) + \bar{m}.$$

Hence, $n + \ell \geqslant m + \ell$. Conversely, by $n + \ell \geqslant m + \ell$, there exists an $\bar{m} \in \bar{\mathbb{N}}$ such that

$$n + \ell = (m + \ell) + \bar{m} = (m + \bar{m}) + \ell.$$

By the cancellation property of addition (see Proposition 1.3), (iii), one obtains (1.17), which means $n \geqslant m$.

(ii) Suppose $\ell > 0$ and $n > m$. Then by Lemma 1.9, (ii)–(iii) and Definition 1.8, $\ell = \bar{\ell} + 1$ for some $\bar{\ell} \in \bar{\mathbb{N}}$, and $n = m + \bar{m}$ for some $\bar{m} \in \mathbb{N}$. Hence,

$$n \times \ell = (m + \bar{m}) \times (\bar{\ell} + 1) = m \times (\bar{\ell} + 1) + \bar{m} \times (\bar{\ell} + 1)$$
$$= m \times \ell + \bar{m} \times \bar{\ell} + \bar{m}.$$

Since $\bar{m} \times \bar{\ell} \in \bar{\mathbb{N}}$ and $\bar{m} > 0$, by Lemma 1.9, (v), one has $\bar{m} \times \bar{\ell} + \bar{m} > 0$. Therefore, by Definition 1.8, $n \times \ell > m \times \ell$.

Conversely, let $n \times \ell > m \times \ell$, with $\ell > 0$. We claim that $n > m$. Otherwise, by Proposition 1.11, one has $n \leqslant m$. Then there exists an $\bar{n} \in \overline{\mathbb{N}}$ such that $m = n + \bar{n}$. Hence,

$$m \times \ell = (n + \bar{n}) \times \ell = n \times \ell + \bar{n} \times \ell.$$

This implies $n \times \ell \leqslant m \times \ell$, contradicting $n \times \ell > m \times \ell$.

(iii) If instead, $n > m$, then by (ii), we have

$$n \times \ell > m \times \ell,$$

a contradiction. Likewise, $n < m$ is also not possible. Hence, we must have $n = m$. The other direction is obvious. $\qquad\square$

Next, we look at another interesting result on $\overline{\mathbb{N}}$.

Proposition 1.13. (Euclidean Algorithm) *Let* $n \in \overline{\mathbb{N}}$, $q \in \mathbb{N}$. *Then there exist* $m, r \in \overline{\mathbb{N}}$ *such that*

$$n = m \times q + r, \qquad 0 \leqslant r < q. \tag{1.18}$$

Proof. Fix $q > 0$, use induction on n.

For $n = 0$: We take $m = 0$ and $r = 0$. Thus, (1.18) holds.

Suppose for n the following holds:

$$n = m \times q + r, \qquad \text{for some } 0 \leqslant r < q.$$

Then for $n + 1$, we have two cases:

(i) $q = r + 1$. For this case,

$$n + 1 = (m \times q + r) + 1 = m \times q + q = (m + 1) \times q.$$

(ii) $q > r + 1$. For this case,

$$n + 1 = (m \times q + r) + 1 = m \times q + (r + 1).$$

Hence, we obtain the result. $\qquad\square$

2 Integers and Rationals

2.1 *Integers*

Recall that on the set $\overline{\mathbb{N}}$, we have a binary operation "+" which maps every given pair (n, m) to a number $n + m \in \overline{\mathbb{N}}$. Interestingly, 0 is a special element in $\overline{\mathbb{N}}$ that has the property

$$n + 0 = 0 + n = n, \qquad \forall n \in \overline{\mathbb{N}}.$$

We call 0 the *unit of addition* in $\bar{\mathbb{N}}$. Further, for $n \geqslant m$, one can find an $\ell \in \bar{\mathbb{N}}$ so that $n = m + \ell$. However, this cannot be done if $m > n$. In particular, for $m > 0$, one could not find an $\ell \in \bar{\mathbb{N}}$ such that

$$0 = m + \ell.$$

Therefore, we would like to find a bigger set containing $\bar{\mathbb{N}}$ so that the above is possible. To this end, let us introduce the following set

$$\bar{\mathbb{N}} \times \bar{\mathbb{N}} = \{(m, n) \mid m, n \in \bar{\mathbb{N}}\},$$

which is called the *Cartesian product* of $\bar{\mathbb{N}}$ with itself. On $\bar{\mathbb{N}} \times \bar{\mathbb{N}}$, we introduce a relation $\sim$ as follows:

$$(m, n) \sim (m', n') \quad \Longleftrightarrow \quad m + n' = m' + n. \tag{2.1}$$

We have the following result concerning this relation.

Proposition 2.1. *The relation $\sim$ is an equivalence relation on $\bar{\mathbb{N}} \times \bar{\mathbb{N}}$, i.e.,*

(i) *(Reflexivity) $(m, n) \sim (m, n)$ for any $(m, n) \in \bar{\mathbb{N}} \times \bar{\mathbb{N}}$;*

(ii) *(Symmetry) $(m, n) \sim (m', n')$ if and only if $(m', n') \sim (m, n)$;*

(iii) *(Transitivity) $(m, n) \sim (m', n')$ and $(m', n') \sim (m'', n'')$ implies that $(m, n) \sim (m'', n'')$.*

Proof. (i) For any $m, n \in \bar{\mathbb{N}}$, by the reflexivity of "$=$" in $\bar{\mathbb{N}}$, we have

$$m + n = m + n.$$

This leads to

$$(m, n) \sim (m, n).$$

(ii) If $(m, n) \sim (m', n')$, then

$$m + n' = m' + n,$$

which, by the symmetry of "$=$" in $\bar{\mathbb{N}}$, is the same as

$$m' + n = m + n'.$$

Hence, $(m', n') \sim (m, n)$.

(iii) Suppose $(m, n) \sim (m', n')$ and $(m', n') \sim (m'', n'')$. Then

$$m + n' = m' + n, \qquad m' + n'' = m'' + n'.$$

This leads to

$$m + n' + m' + n'' = m' + n + m'' + n'.$$

By the cancellation property of addition (Proposition 1.3, (iii)), we obtain

$$m + n'' = m'' + n.$$

Thus, $(m, n) \sim (m'', n'')$.   □

Having the above, we now introduce the following definition.

Definition 2.2. For any $(m, n) \in \overline{\mathbb{N}} \times \overline{\mathbb{N}}$, let

$$m \ominus n = \big\{ (m', n') \in \overline{\mathbb{N}} \times \overline{\mathbb{N}} \mid (m', n') \sim (m, n) \big\}, \tag{2.2}$$

which is called the *equivalent class* of (m, n) under the equivalent relation $\sim$ defined by (2.1). Any element $(m', n') \in m \ominus n$ is called a *representative* of its equivalent class. Let

$$\mathbb{Z} = \big\{ m \ominus n \mid (m, n) \in \overline{\mathbb{N}} \times \overline{\mathbb{N}} \big\}.$$

Any element $m \ominus n \in \mathbb{Z}$ is called an *integer*.

Lemma 2.3. *Let* $m, n, m', n' \in \overline{\mathbb{N}}$. *Then*

$$(m, n) \sim (m', n') \quad \Longleftrightarrow \quad m \ominus n = m' \ominus n'. \tag{2.3}$$

In particular, for any $m, n, \ell \in \overline{\mathbb{N}}$,

$$(m + \ell) \ominus (n + \ell) = m \ominus n. \tag{2.4}$$

Proof. Necessity. For any $(m'', n'') \in m \ominus n$, we have

$$(m'', n'') \sim (m, n) \sim (m', n').$$

Thus, $(m'', n'') \in m' \ominus n'$, leading to

$$m \ominus n \subseteq m' \ominus n'.$$

The other inclusion direction is similar.

Sufficiency. Since

$$(m, n) \in m \ominus n = m' \ominus n',$$

we obtain $(m, n) \sim (m', n')$.

The last conclusion is obvious.   □

Let

$$\mathbb{Z}_0 = \{ m \ominus 0 \mid m \in \overline{\mathbb{N}} \} \subset \mathbb{Z}.$$

If we define a map $\varphi : \overline{\mathbb{N}} \to \mathbb{Z}$ by

$$\varphi(m) = m \ominus 0, \qquad \forall m \in \overline{\mathbb{N}},$$

then the image $\varphi(\overline{\mathbb{N}}) = \mathbb{Z}_0$. Moreover, $\varphi(m) \neq \varphi(m')$ if and only if $m \neq m'$. Hence, $\overline{\mathbb{N}}$ can be identified as $\mathbb{Z}_0$, a subset of $\mathbb{Z}$. Because of this, hereafter, we directly write

$$n \ominus 0 = n, \qquad \forall n \in \overline{\mathbb{N}}. \tag{2.5}$$

In particular, one has

$$(0 \ominus 0) = 0, \qquad (1 \ominus 0) = 1.$$

Next, we extend the addition and multiplication from $\overline{\mathbb{N}}$ to $\mathbb{Z}$.

Definition 2.4. For any $(m \ominus n), (p \ominus q) \in \mathbb{Z}$, define addition and multiplication of them as follows:

$$(m \ominus n) + (p \ominus q) = (m + p) \ominus (n + q),$$
$$(m \ominus n) \times (p \ominus q) = (m \times p + n \times q) \ominus (m \times q + p \times n).$$

The following proposition is necessary before going further.

Proposition 2.5. *Addition and multiplication of integers are well-defined.*

Proof. Let

$$m \ominus n = m' \ominus n', \qquad p \ominus q = p' \ominus q',$$

which means $(m, n) \sim (m', n')$ and $(p, q) \sim (p', q')$. Thus,

$$m + n' = m' + n, \qquad p + q' = p' + q.$$

Add the above together, we have

$$(m + n') + (p + q') = (m' + n) + (p' + q),$$

which is the same as

$$(m + p) + (n' + q') = (m' + p') + (n + q).$$

Hence,

$$(m + p) \ominus (n + q) = (m' + p') \ominus (n' + q').$$

By definition, this means

$$(m \ominus n) + (p \ominus q) = (m' \ominus n') + (p' \ominus q').$$

Therefore, the extended addition is well-defined on $\mathbb{Z}$.

Now, for multiplication, we show the following: If

$$m \ominus n = m' \ominus n',$$

then for any $p \ominus q \in \mathbb{Z}$,

$$(m \ominus n) \times (p \ominus q) = (m' \ominus n') \times (p \ominus q). \tag{2.6}$$

This is equivalent to the following:

$$(m \times p + n \times q) \ominus (m \times q + n \times p)$$
$$= (m' \times p + n' \times q) \ominus (m' \times q + n' \times p),$$

which is equivalent to

$$m \times p + n \times q + m' \times q + n' \times p = m' \times p + n' \times q + m \times q + n \times p.$$

By the rules of addition and multiplication on $\overline{\mathbb{N}}$, we have

$$(m + n') \times p + (m' + n) \times q = (m' + n) \times p + (m + n') \times q.$$

Since $m + n' = m' + n$, the above holds true which means that (2.6) holds. Therefore, the extended multiplication is well-defined on $\mathbb{Z}$. $\qquad\square$

Let us look at the following special cases:

$$\begin{aligned} (n \ominus 0) + (m \ominus 0) &= (n + m) \ominus 0, \\ (n \ominus 0) \times (m \ominus 0) &= (n \times m) \ominus 0, \end{aligned} \qquad \forall n, m \in \overline{\mathbb{N}}.$$

Hence, besides that $\overline{\mathbb{N}}$ can be identified as the subset $\mathbb{Z}_0$ of $\mathbb{Z}$, the additions and multiplications defined on $\overline{\mathbb{N}}$ and $\mathbb{Z}$ are compatible.

The following definition gives a new feature of the integer set $\mathbb{Z}$ that the set $\overline{\mathbb{N}}$ does not have.

Definition 2.6. (Negation of integers) For any $(m \ominus n) \in \mathbb{Z}$, define

$$-(m \ominus n) = n \ominus m,$$

which is called the *negation* of $(m \ominus n)$. In particular, if $n = n \ominus 0$ with $n \in \overline{\mathbb{N}}$ (which is also called a *non-negative integer* from now on), we denote

$$-(n \ominus 0) = (0 \ominus n) = -n,$$

and call it a *non-positive integer*. If $a \in \mathbb{Z}$ is non-positive and is different from 0, we call it a *negative integer*, denoted by $a < 0$. Any natural number is called a *positive integer*.

We now collect some laws of addition and multiplication for integers.

Proposition 2.7. (Algebraic Laws for Integers) *Let $a, b, c \in \mathbb{Z}$. Then*

$$a + b = b + a,$$
$$(a + b) + c = a + (b + c),$$
$$a + 0 = 0 + a = a,$$
$$a + (-a) = (-a) + a = 0,$$
$$a \times b = b \times a,$$
$$(a \times b) \times c = a \times (b \times c),$$
$$a \times 1 = 1 \times a = a,$$
$$a \times (b + c) = a \times b + a \times c.$$

Proof. To prove the first, we write $a = m \ominus n$, $b = p \ominus q$. Then use definition of addition, we have

$$a + b = (m \ominus n) + (p \ominus q) = (m + p) \ominus (n + q)$$
$$= (p + m) \ominus (q + n) = (p \ominus q) + (m \ominus n) = b + a.$$

The proofs of others are similar, which are left to the readers. □

According to the above, we have

$$(0 \ominus 1) = -1,$$

and

$$(-1) \times n = n \times (-1) = (n \ominus 0) \times (0 \ominus 1)$$
$$= (n \times 0 + 0 \times 1) \ominus (n \times 1 + 0 \times 0) = (0 \ominus n) = -n.$$

Further, for any $m \ominus n \in \mathbb{Z}$, we have

$$m \ominus n = \begin{cases} (m - n) \ominus 0 = m - n, & m \geqslant n, \\ 0 \ominus (n - m) = -(n - m), & m < n. \end{cases} \tag{2.7}$$

Hence, the following representation holds:

$$\mathbb{Z} = \{\cdots, -3, -2, -1, 0, 1, 2, 3, \cdots\}. \tag{2.8}$$

We now define *subtraction* on $\mathbb{Z}$ by the following:

$$a - b = a + (-b), \qquad \forall a, b \in \mathbb{Z},$$

and call $a - b$ the *difference* between a and b. Note that if $a = m \ominus n$ and $b = p \ominus q$ for some $m, n, p, q \in \overline{\mathbb{N}}$, then

$$a - b = (m \ominus n) - (p \ominus q) = (m \ominus n) + (q \ominus p) = (m + q) \ominus (n + p).$$

The following proposition collects some more laws for integers.

Proposition 2.8. *Let $a, b, c \in \mathbb{Z}$, with $c \neq 0$. Then*

$$a - b = 0 \quad \Longleftrightarrow \quad a = b. \tag{2.9}$$

$$(-a) \times b = a \times (-b) = -(a \times b). \tag{2.10}$$

$$a \times b = 0 \quad \Longleftrightarrow \quad a = 0, \ \text{or} \ b = 0. \tag{2.11}$$

$$a \times c = b \times c \quad \Longleftrightarrow \quad a = b. \tag{2.12}$$

Proof. Let $n, m, p, q \in \overline{\mathbb{N}}$ such that

$$a = m \ominus n, \quad b = p \ominus q.$$

For (2.9), we look at the following:

$$0 \ominus 0 = 0 = a - b = (m \ominus n) - (p \ominus q) = (m \ominus n) + (-(p \ominus q))$$
$$= (m \ominus n) + (q \ominus p) = (m + q) \ominus (n + p).$$

This is equivalent to

$$m + q = n + p,$$

from which, one has

$$a = m \ominus n = p \ominus q = b.$$

Hence, (2.9) holds. Next, for (2.10), we observe the following:

$$(-a) \times b = (-(m \ominus n)) \times (p \ominus q) = (n \ominus m) \times (p \ominus q)$$
$$= (n \times p + m \times q) \ominus (m \times p + n \times q) = (m \ominus n) \times (q \ominus p) = a \times (-b).$$

Also we note that

$$-(a \times b) = -((m \ominus n) \times (p \ominus q))$$
$$= -((m \times p + n \times q) \ominus (n \times p + m \times q))$$
$$= (n \times p + m \times q) \ominus (m \times p + n \times q)$$
$$= (n \ominus m) \times (p \ominus q) = (-a) \times b.$$

This proves (2.10).

Now, we look at (2.11). Suppose both a and b are non-zero and $a \times b = 0$. Then we have the following cases:

- $a, b > 0$. By Proposition 1.12, (ii), $a \times b > 0$, a contradiction.

• $a > 0$ and $b < 0$. Then $-b > 0$ and by Proposition 1.12, together with (2.10), one has

$$0 < a \times (-b) = -(a \times b).$$

This implies $a \times b \neq 0$, a contradiction.

Other two cases (i.e., $a < 0$, $b > 0$; and $a < 0$, $b < 0$) can be treated similarly. Thus, the necessity of (2.11) holds. The sufficiency is obvious.

We now prove (2.12). If $a \times c = b \times c$ and $c \neq 0$, then we claim that $a - b = 0$. If not, then

$$0 \neq (a - b) \times c = (a + (-b)) \times c = a \times c + (-b) \times c$$
$$= a \times c - (b \times c).$$

Hence, by (2.9), $a \times c \neq b \times c$, a contradiction. Consequently, $a - b = 0$. Thus, the necessity of (2.12) holds. The sufficiency is obvious. $\qquad \square$

Note that for any $a, b \in \overline{\mathbb{N}}$,

$$a - b = a + (-b) = (a \ominus 0) + (0 \ominus b) = a \ominus b.$$

Further, for any $a, b \in \overline{\mathbb{N}}$ with $a \geqslant b$, we have

$$a = b + c,$$

for some $c \in \overline{\mathbb{N}}$, and

$$a - b = (b + c) \ominus b = c \ominus 0 = c.$$

Because of such a compatibility, from now on, we replace $\ominus$ by $-$, for notational simplicity.

Now, we extend the orders ">" and $\geqslant$" from $\overline{\mathbb{N}}$ to $\mathbb{Z}$.

Definition 2.9. Let $a, b \in \mathbb{Z}$. We say that $a > b$ (or $b < a$) if $a - b \in \mathbb{N}$, and $a \geqslant b$ (or $b \leqslant a$) if $a - b \in \overline{\mathbb{N}}$.

The following proposition is concerned with the order on $\mathbb{Z}$, whose proof is left to the readers.

Proposition 2.10. *Let* $a, b, c \in \mathbb{Z}$.

(i) (Addition preserves order) $a > b$ *if and only if* $a + c > b + c$.

(ii) (Positive multiplication preserves order) *Let* $c > 0$. *Then* $a > b$ *if and only if* $a \times c > b \times c$.

(iii) (Negation reverses order) $a > b$ *if and only if* $-a < -b$.

(iv) (Order is transitive) *If* $a > b$ *and* $b > c$, *then* $a > c$.

(v) (Order trichotomy) *Exactly one of* $a > b$, $a < b$, *and* $a = b$ *holds*.

To conclude this subsection, let us point out one fact: The principle of induction does not apply to the integers. Namely, if $P(n)$ is a statement pertaining to $n \in \mathbb{Z}$, if $P(0)$ holds and whenever $P(n)$ holds, $P(n+1)$ holds. This does not imply that $P(n)$ holds for all $n \in \mathbb{Z}$. Here is a simple example: Let $P(n)$ be the following statement: "n is non-negative". Clearly, $P(0)$ is true and when $P(n)$ holds, $P(n+1)$ holds. But, "$P(n)$ is true for all $n \in \mathbb{Z}$" is not true.

2.2 Rational numbers

Note that for any $a, b \in \mathbb{Z}$, both $a + b$ and $a - b$ are in $\mathbb{Z}$. Such a property is described by saying that $\mathbb{Z}$ is closed under addition and subtraction. On the other hand, for any $a, b \in \mathbb{Z}$, $a \times b \in \mathbb{Z}$, and integer 1 has an interesting property that

$$1 \times a = a \times 1 = a, \qquad \forall a \in \mathbb{Z}.$$

We call 1 the *unit of multiplication* in $\mathbb{Z}$. Now, for any positive integers a, b with $a > b$, sometimes one can find an integer c such that $a = b \times c$. In this case, we say that a is a *multiple* of b, b is a *divisor* of a, and a is *divisible* by b; for example $a = 6$, $b = 2$, then $c = 3$. But sometimes one could not do that, for example, $a = 5$, $b = 2$, i.e., there exists no $c \in \mathbb{Z}$ such that $5 = 2 \times c$. Therefore, we would like to further extend the set $\mathbb{Z}$. To this end, we introduce a relation on the set $\mathbb{Z} \times (\mathbb{Z} \setminus \{0\})$ as follows:

$$(a, b) \sim (c, d) \qquad \text{if} \qquad a \times d = c \times b. \tag{2.13}$$

We claim that $\sim$ is an equivalent relation, i.e., the following hold:

Reflexivity: $(a, b) \sim (a, b)$, since $a \times b = a \times b$.

Symmetry: $(a, b) \sim (c, d)$ implies $(c, d) \sim (a, b)$ since

$$a \times d = c \times b \quad \Rightarrow \quad c \times b = a \times d.$$

Transitivity: $(a, b) \sim (c, d)$, $(c, d) \sim (e, f)$ implies $(a, b) \sim (e, f)$, since

$$a \times d = c \times b, \quad c \times f = e \times d,$$

imply

$$a \times d \times c \times f = c \times b \times e \times d.$$

Since $d \neq 0$, by cancellation property, the above implies

$$a \times c \times f = c \times b \times e.$$

This leads to the following: either $c = 0$, or $a \times f = e \times b$. In the case, $c = 0$, we must have $a = e = 0$ since $d \neq 0$. Hence, in either case, we always have

$$a \times f = e \times b,$$

which gives $(a, b) \sim (e, f)$.

Now, for any $(a, b) \in \mathbb{Z} \times (\mathbb{Z} \setminus \{0\})$, we define corresponding equivalence class

$$a/b = \{(c, d) \in \mathbb{Z} \times (\mathbb{Z} \setminus \{0\}) \mid (c, d) \sim (a, b)\},$$

and let

$$\mathbb{Q} = \Big\{a/b \mid (a, b) \in \mathbb{Z} \times (\mathbb{Z} \setminus \{0\})\Big\}.$$

Any element in $\mathbb{Q}$ is called a *rational number*. Next, we extend the addition, multiplication and negation from $\mathbb{Z}$ to $\mathbb{Q}$.

Definition 2.11. For any $a/b, c/d \in \mathbb{Q}$, we define the addition, multiplication, and negation as follows:

$$(a/b) + (c/d) = (a \times d + b \times c)/(b \times d),$$
$$(a/b) \times (c/d) = (a \times c)/(b \times d),$$
$$-(a/b) = (-a/b).$$

Proposition 2.12. *The above defined operations are well-defined.*

Proof. We show the conclusion for the addition. Let $a/b, a'/b', c/d \in \mathbb{Q}$ with $a/b = a'/b'$. Then

$$a \times b' = a' \times b.$$

It suffices to show that

$$(a \times d + c \times b)/(b \times d) = (a/b) + (c/d)$$
$$= (a'/b') + (c/d) = (a' \times d + c \times b')/(b' \times d).$$

Or equivalently,

$$(a \times d + c \times b) \times (b' \times d) = (a' \times d + c \times b') \times (b \times d).$$

This is the same as (making use of Proposition 2.7)

$$a \times b' \times d \times d + c \times b \times b' \times d = a' \times b \times d \times d + c \times b \times b' \times d,$$

which is equivalent to

$$a \times b' \times d \times d = a' \times b \times d \times d.$$

Since $d \neq 0$, the above is equivalent to

$$a \times b' = a' \times b,$$

which is true.

The proofs of the other two are similar.   □

The following result collects some interesting facts that reveals the relation between $\mathbb{Z}$ and $\mathbb{Q}$.

Proposition 2.13. *For any $a, b \in \mathbb{Z}$,*

$$(a/1) + (b/1) = (a+b)/1,$$
$$(a/1) \times (b/1) = (a \times b)/1,$$
$$-(a/1) = (-a)/1,$$

and in the case $b \neq 0$,

$$(b/b) = (1/1), \quad (0/b) = (0/1),$$
$$(a/b) = (1/1) \quad \Longleftrightarrow \quad a = b,$$
$$(a/b) = (0/1) \quad \Longleftrightarrow \quad a = 0,$$
$$(a/b) + (0/1) = a/b,$$
$$(a/b) \times (0/1) = (0/1).$$

Proof. It is clear that

$$(a/1) + (b/1) = (a \times 1 + b \times 1)/(1 \times 1) = (a+b)/1.$$

The others are similar.   □

Because of the above, from now on, we identify

$$a/1 = a, \qquad \forall a \in \mathbb{Z}. \tag{2.14}$$

In particular,

$$0/1 = 0, \qquad 1/1 = 1. \tag{2.15}$$

Hence, from (2.14), $\mathbb{Z}$ becomes a subset of $\mathbb{Q}$:

$$\mathbb{Z} \subset \mathbb{Q}. \tag{2.16}$$

For any $a, b \in \mathbb{Z} \setminus \{0\}$, $a/b \neq 0$. In this case, we define the *reciprocal* of a/b by

$$(a/b)^{-1} = b/a.$$

One can check that this operation is also well-defined. Note that 0 does not have its reciprocal.

Proposition 2.14. (Algebraic Laws for Rationals) *Let $\xi, \eta, \zeta \in \mathbb{Q}$. Then*

$$\xi + \eta = \eta + \xi,$$
$$(\xi + \eta) + \zeta = \xi + (\eta + \zeta),$$
$$\xi + 0 = 0 + \xi = \xi,$$
$$\xi + (-\xi) = (-\xi) + \xi = 0,$$
$$\xi \times \eta = \eta \times \xi,$$
$$(\xi \times \eta) \times \zeta = \xi \times (\eta \times \zeta),$$
$$\xi \times 1 = 1 \times \xi = \xi,$$
$$\xi \times (\eta + \zeta) = \xi \times \eta + \xi \times \zeta.$$

If $\xi \neq 0$, then ξ^{-1} is defined and non-zero as well. Moreover,

$$\xi \times \xi^{-1} = \xi^{-1} \times \xi = 1.$$

Proof. We prove the second equality. For $\xi, \eta, \zeta \in \mathbb{Q}$, let

$$\xi = a/b, \quad \eta = c/d, \quad \zeta = e/f, \qquad a, c, e \in \mathbb{Z}, \; b, d, f \in \mathbb{Z} \setminus \{0\}.$$

Then

$$(\xi + \eta) + \zeta = \big((a/b) + (c/d)\big) + (e/f)$$
$$= \big((a \times d + b \times c)/(b \times d)\big) + (e/f)$$
$$= \big((a \times d + b \times c) \times f + b \times d \times e\big)/(b \times d \times f),$$

and

$$\xi + (\eta + \zeta) = (a/b) + \big((c/d) + (e/f)\big)$$
$$= (a/b) + \big((c \times f + d \times e)/(d \times f)\big)$$
$$= \big(a \times d \times f + b \times (c \times f + d \times e)\big)/(b \times d \times f).$$

Hence,

$$(\xi + \eta) + \zeta = \xi + (\eta + \zeta).$$

The others can be proved similarly. □

For notational simplicity, from now on, for any $\xi, \eta \in \mathbb{Q}$, we write

$$\xi \times \eta = \xi \cdot \eta = \xi\eta,$$

unless the symbol "$\times$" or "$\cdot$" needs to be emphasized.

Now, for any $\xi, \eta \in \mathbb{Q}$ with $\eta \neq 0$, we define the *quotient* ξ/η as follows:

$$\xi/\eta = \xi\eta^{-1}. \tag{2.17}$$

The above is also denoted by $\xi \div \eta$, or $\frac{\xi}{\eta}$. The operation $\div$ is called the *division*. Clearly, if $\zeta = \xi \cdot \eta$, then

$$\xi = \zeta \div \eta = \zeta/\eta = \frac{\zeta}{\eta}, \qquad \eta = \zeta \div \xi = \zeta/\xi = \frac{\zeta}{\xi}.$$

The following extends the order from $\mathbb{Z}$ to $\mathbb{Q}$.

Definition 2.15. (i) $\xi \in \mathbb{Q}$ is *positive*, denoted by $\xi > 0$, if there are $a, b \in \mathbb{N}$ such that $\xi = a/b$; $\xi \in \mathbb{Q}$ is *negative*, denoted by $\xi < 0$, if $-\xi$ is positive; $\xi \in \mathbb{Q}$ is *non-negative*, denoted by $\xi \geqslant 0$, if $\xi > 0$ or $\xi = 0$; $\xi \in \mathbb{Q}$ is *non-positive*, denoted by $\xi \leqslant 0$, if $-\xi$ is non-negative.

(ii) Let $\xi, \eta \in \mathbb{Q}$. We say that $\xi > \eta$ if $\xi - \eta > 0$, and $\xi \geqslant \eta$ if $\xi - \eta \geqslant 0$.

We have the following result which is comparable with that for integers (see Proposition 2.10).

Proposition 2.16. *Let* $\xi, \eta, \zeta \in \mathbb{Q}$. *Then*

(i) (Addition preserves order) $\xi > \eta$ *if and only if* $\xi + \zeta > \eta + \zeta$.

(ii) (Positive multiplication preserves order) *Let* $\zeta > 0$. *Then* $\xi > \eta$ *if and only if* $\xi \cdot \zeta > \eta \cdot \zeta$.

(iii) (Negation reverses order) $\xi > \eta$ *if and only if* $-\xi < -\eta$.

(iv) (Order is transitive) *If* $\xi > \eta$ *and* $\eta > \zeta$, *then* $\xi > \zeta$.

(v) (Trichotomy of rationals) *Exactly one of the three statements holds:* $\xi < \eta$, $\xi = \eta$, *and* $\xi > \eta$.

(vi) *If* $\xi > \eta > 0$, *then* $0 < \xi^{-1} < \eta^{-1}$.

Proof. We only prove (vi), leaving the others to the readers. By $\xi > \eta > 0$, we may assume that

$$\xi = a/b, \quad \eta = c/d,$$

for some $a, b, c, d \in \mathbb{N}$, and

$$0 < \xi - \eta = a/b - c/d = a/b + (-c)/d$$
$$= [ad + (-c)b]/bd = (ad - bc)/bd.$$

Thus $ad - bc > 0$. Now, we look at

$$\eta^{-1} - \xi^{-1} = d/c - b/a = (ad - bc)/ac > 0.$$

It is obvious that $\xi^{-1} > 0$. This proves (vi). $\qquad\square$

Next, we introduce the following notion.

Definition 2.17. (i) For any $\xi \in \mathbb{Q}$, define its *absolute value* $|\xi|$ by the following:

$$|\xi| = \begin{cases} \xi, & \text{if } \xi \geqslant 0, \\ -\xi, & \text{if } \xi < 0. \end{cases}$$

(ii) For any $\xi, \eta \in \mathbb{Q}$, define the *distance* between ξ and η by the following:

$$d(\xi, \eta) = |\xi - \eta|.$$

The following result is concerned with the absolute value and distance. The proof is left to the readers.

Proposition 2.18. *Let $\xi, \eta, \zeta \in \mathbb{Q}$. Then*

(i) $|\xi| \geqslant 0$ *and* $|\xi| = 0$ *if and only if* $\xi = 0$.

(ii) $|\xi + \eta| \leqslant |\xi| + |\eta|$.

(iii) *For* $\eta \geqslant 0$, $-\eta \leqslant \xi \leqslant \eta$ *if and only if* $|\xi| \leqslant \eta$. *In particular,* $-|\xi| \leqslant \xi \leqslant |\xi|$.

(iv) $|\xi \cdot \eta| = |\xi| \cdot |\eta|$. *In particular,* $|-\xi| = |\xi|$.

(v) $d(\xi, \eta) \geqslant 0$ *and* $d(\xi, \eta) = 0$ *if and only if* $\xi = \eta$.

(vi) $d(\xi, \eta) = d(\eta, \xi)$.

(vii) $d(\xi, \zeta) \leqslant d(\xi, \eta) + d(\eta, \zeta)$.

Next, we introduce the following notion called the *exponentiation*.

Definition 2.19. (i) For any $a \in \mathbb{N}$, define

$$0^a = 0.$$

(ii) Let $\xi \in \mathbb{Q} \setminus \{0\}$. Define

$$\xi^0 = 1, \quad \xi^{n+1} = \xi^n \cdot \xi, \quad \forall n \in \mathbb{N},$$

and define

$$\xi^{-n} = 1/\xi^n, \quad \forall n \in \mathbb{N}.$$

We have the following laws for the exponentiation.

Proposition 2.20. *Let $\xi, \eta \in \mathbb{Q} \setminus \{0\}$.*

(i) *For any $a, b \in \mathbb{Z}$, the following hold:*

$$\xi^a \cdot \xi^b = \xi^{a+b}, \quad (\xi^a)^b = \xi^{ab}, \quad (\xi \cdot \eta)^a = \xi^a \cdot \eta^a, \quad |\xi^a| = |\xi|^a.$$

(ii) *For $a \in \mathbb{Z} \setminus \{0\}$, $\xi^a = \eta^a$ if and only if $\xi = \eta$.*

(iii) *For $n \in \mathbb{N}$,*

$$\xi > \eta > 0 \quad \Rightarrow \quad \xi^n > \eta^n > 0, \quad \eta^{-n} > \xi^{-n} > 0.$$

Proof. (i) For the first equality, we first show that $\xi^m \cdot \xi^n = \xi^{m+n}$ for all $\xi \neq 0$ and $m, n \in \overline{\mathbb{N}}$. To this end, we let $m \in \overline{\mathbb{N}}$ be fixed and use induction in n.

For $n = 0$: $\xi^m \cdot \xi^0 = \xi^m \cdot 1 = \xi^m = \xi^{m+0}$.

Suppose the conclusion holds for n. Then for $n + 1$, by definition and induction hypothesis, we have

$$\xi^m \cdot \xi^{n+1} = \xi^m \cdot \xi^n \cdot \xi = \xi^{m+n} \cdot \xi = \xi^{m+n+1}.$$

This means that $\xi^a \cdot \xi^b = \xi^{a+b}$ holds for $a, b \in \overline{\mathbb{N}}$.

Now, if $a = -n$, $b = -m$ with $n, m \in \overline{\mathbb{N}}$, then

$$\xi^a \cdot \xi^b = \xi^{-n} \cdot \xi^{-m} = (1/\xi^n) \cdot (1/\xi^m) = 1/(\xi^n \cdot \xi^m)$$
$$= 1/\xi^{n+m} = \xi^{-(n+m)} = \xi^{a+b}.$$

Next, if $a = m$ and $b = -n$, then we may let, say, $m > n$. Consequently,

$$\xi^a \cdot \xi^b = \xi^m \cdot \xi^{-n} = \xi^{m-n} \cdot \xi^n \cdot 1/\xi^n = \xi^{m-n} = \xi^{a+b}.$$

The case $m \leqslant n$ can be treated similarly. Hence, the first equality holds.

For the second equality, first let $a \in \mathbb{Z}$ and $b = n \in \overline{\mathbb{N}}$ (still let $\xi \neq 0$). Then we use induction on n. The case $n = 0$ is trivial. Suppose the equality holds for n. Then for $n + 1$, we have, making use of the first equality,

$$(\xi^a)^{n+1} = (\xi^a)^n \cdot (\xi^a) = \xi^{an} \cdot \xi^a = \xi^{an+a} = \xi^{a(n+1)}.$$

This completes the induction. In the case that $b = -n$ with $n \in \overline{\mathbb{N}}$, we have

$$(\xi^a)^b = (\xi^a)^{-n} = 1/(\xi^a)^n = 1/\xi^{an} = \xi^{-an} = \xi^{ab}.$$

For the third equality, we first let $a = n \in \overline{\mathbb{N}}$. Use induction on n. The case $n = 0$ is trivial. Suppose the statement is true for n. Then for $n + 1$, we have (using the commutativity of multiplication)

$$(\xi \cdot \eta)^{n+1} = (\xi \cdot \eta)^n \cdot (\xi \cdot \eta) = \xi^n \cdot \eta^n \cdot \xi \cdot \eta = \xi^{n+1} \cdot \eta^{n+1}.$$

This proves the third equality for the case that $a = n \in \overline{\mathbb{N}}$. Next, let $a = -n$ with $n \in \overline{\mathbb{N}}$. Then by what we have proved

$$(\xi \cdot \eta)^{-n} = 1/(\xi \cdot \eta)^n = 1/(\xi^n \cdot \eta^n) = 1/\xi^n \cdot 1/\eta^n = \xi^{-n} \cdot \eta^{-n}.$$

Hence, the third equality holds.

For the fourth equality, if $\xi \geqslant 0$, it is trivial. Let now $\xi < 0$. Then $\xi = -|\xi| = (-1) \cdot |\xi|$. Hence, by the proved third equality,

$$\xi^a = [(-1) \cdot |\xi|]^a = (-1)^a \cdot |\xi|^a.$$

Consequently,

$$|\xi^a| = \left|(-1)^a \cdot |\xi|^a\right| = |\xi|^a.$$

The proofs of the rest of the conclusions are left to the readers. $\qquad\square$

Next, we look at an interesting issue of rational numbers.

Proposition 2.21. (i) (Interspersing integers by rationals) *For any $\xi \in \mathbb{Q}$, there exists a unique integer, denoted by $[\xi]$ such that $[\xi] \leqslant \xi < [\xi] + 1$.*

(ii) (Interspersing of rationals by rationals) *If $\xi, \eta \in \mathbb{Q}$ with $\xi < \eta$. Then there exists a $\zeta \in \mathbb{Q}$ such that $\xi < \zeta < \eta$.*

Proof. (i) Let $\xi = n/q$ with $n, q \in \overline{\mathbb{N}}$, $q > 0$. Then by Euclidean algorithm, there exist $r, m \in \overline{\mathbb{N}}$ with $0 \leqslant r < q$ such that

$$n = mq + r.$$

Hence,

$$n \geqslant mq.$$

Multiplying $q^{-1} > 0$, one gets

$$\xi = n/q = (mq + r)q^{-1} = m + rq^{-1} \geqslant m.$$

Also, since $r < q$, there exists a $k \in \overline{\mathbb{N}}$, $k > 0$, such that

$$q = r + k.$$

Hence,

$$n + k = mq + r + k = (m + 1)q.$$

This gives

$$m + 1 = (n + k)/q = n/q + k/q > n/q = \xi \geqslant m.$$

Now, if $\xi < 0$, then $-\xi > 0$. Hence, from the above, we have some $m \in \overline{\mathbb{N}}$ such that

$$m \leqslant -\xi < m + 1.$$

Consequently,

$$-(m + 1) < \xi \leqslant -m.$$

Hence, we can always find a $p \in \mathbb{Z}$ such that

$$p \leqslant \xi < p + 1.$$

Now, suppose there exists another (different) $\bar{p} \in \mathbb{Z}$ such that

$$\bar{p} \leqslant \xi < \bar{p} + 1.$$

Without loss of generality, let $p < \bar{p}$. Then $p + 1 \leqslant \bar{p}$. Consequently,

$$\xi < p + 1 \leqslant \bar{p} \leqslant \xi,$$

a contradiction. Hence, p is unique, and we denote $[\xi] = p$.

(ii) Note that $1/2 \in \mathbb{Q}$ and it is positive. Hence,

$$\xi/2 < \eta/2.$$

Consequently,

$$\xi = \xi/2 + \xi/2 < \xi/2 + \eta/2 < \eta/2 + \eta/2 = \eta.$$

Clearly, for $\zeta = \xi/2 + \eta/2 \in \mathbb{Q}$, we have $\xi < \zeta < \eta$. $\qquad\square$

For any $(a, b) \in \mathbb{Z} \times (\mathbb{Z} \setminus \{0\})$, we may always find a representative $(a', b') \in a/b$ such that $b' \in \mathbb{N}$. Therefore, hereafter, without loss of generality, we may let

$$\mathbb{Q} = \{m/n \mid m = 0, \pm 1, \pm 2, \cdots, \ n = 1, 2, 3, \cdots\}. \tag{2.18}$$

Let $\mathbb{L}$ be a horizonal straight line on which we define an order as follows: For any $x, y \in \mathbb{L}$,

$$x < y \qquad \text{if and only if } x \text{ is on the left of } y.$$

Fix a point, label it 0 and call it the *origin*. Pick a unit length. For each $n \in \overline{\mathbb{N}}$, find a point on the line $\mathbb{L}$, label it n, which is on the right of the origin with the distance of n units from the origin. Also, find a point, label it $-n$, which is on the left of the origin with the distance of n units from the origin. Then $\mathbb{Z}$ is labeled on $\mathbb{L}$. For any rational number of form m/n with $m \in \overline{\mathbb{N}}$, $n \in \mathbb{N}$, by Euclidean algorithm, we have

$$m = nq + r, \qquad 0 \leqslant r < n, \qquad q \in \overline{\mathbb{N}}.$$

We mark m/n on the line $\mathbb{L}$ as follows: equally divide the line segment of $\mathbb{L}$ between the points q and $q + 1$ into n parts with the partition points:

$$q = \xi_0 < \xi_1 < \xi_2 < \cdots < \xi_n = q + 1.$$

Then mark m/n at ξ_r. For negative rational numbers, we can do the similar thing. As a result, we see that

$$\overline{\mathbb{N}} \subset \mathbb{Z} \subset \mathbb{Q} \subseteq \mathbb{L}, \qquad (2.19)$$

We call the points on $\mathbb{L}$ corresponding to those in $\mathbb{N}$, $\mathbb{Z}$, and $\mathbb{Q}$ the *natural number*, the *integer*, and the *rational number points*, respectively. By Proposition 2.21, $\mathbb{Q}$ is dense in $\mathbb{L}$, in the sense that between every two rational number points, there exists a rational number point. A natural question is: Do rational numbers occupy the whole line $\mathbb{L}$? Equivalently, is the following true:

$$\mathbb{Q} = \mathbb{L}?$$

The following result answers the question negatively.

Proposition 2.22. *There does not exist a rational number $\xi \in \mathbb{Q}$ for which $\xi^2 = 2$. This means that the side length of a square with area 2 is not a rational number.*

Proof. Suppose there exists a rational number $\xi = p/q$ with $p, q \in \mathbb{N}$, $q \neq 0$, and they are *co-prime*, i.e., the greatest common divisor of p and q is 1, such that

$$2 = \xi^2 = p^2/q^2.$$

This implies

$$2q^2 = p^2.$$

Since p and q are co-prime, the above tells us that p is divisible by 2. Hence, we may write $p = 2\bar{p}$. Then the above leads to

$$q^2 = 2\bar{p}^2.$$

Similar to the above argument, we must have that q is divisible by 2. Hence, both p and q are divisible by 2, contradicting our assumption. □

The above result shows that $\mathbb{L} \setminus \mathbb{Q} \neq \varnothing$. We call any $x \in \mathbb{L} \setminus \mathbb{Q}$ an *irrational point*.

3   Real Numbers

We have seen that although the set $\mathbb{Q}$ of rational numbers is closed under addition, subtraction, multiplication, and division, there is still a big unsatisfactory point: $\mathbb{Q} \neq \mathbb{L}$. In another word, there are (a lot of) points on the line $\mathbb{L}$ that are not rational points. Therefore, we want to further extend the set $\mathbb{Q}$ of rational numbers.

3.1   *Sequences of rational numbers*

Definition 3.1. A set of ordered rational numbers $\xi_1, \xi_2, \xi_3, \cdots$ is called a *sequence* of rational numbers, denoted by $\{\xi_n\}_{n \geqslant 1}$. The set of all sequences of rational numbers is denoted by $\ell^0(\mathbb{Q})$.

• For any sequence $\{\xi_n\}_{n \geqslant 1} \in \ell^0(\mathbb{Q})$, it is allowed to have $\xi_n = \xi_m$ for some $n \neq m$.

• Sequence could be a finite one. For convenience hereafter, if $\xi_1, \xi_2, \cdots, \xi_k$ is a finite sequence of rational numbers, we define $\xi_n = \xi_k$, for all $n > k$. Therefore, we will only consider infinite sequences.

Definition 3.2. (i) A sequence $\{\xi_n\}_{n \geqslant 1} \in \ell^0(\mathbb{Q})$ is said to be *bounded* if there exists a rational number $M > 0$ such that

$$|\xi_n| \leqslant M, \qquad \forall n \geqslant 0.$$

The set of all bounded sequences of rational numbers is denoted by $\ell^\infty(\mathbb{Q})$.

(ii) A sequence $\{\xi_n\}_{n \geqslant 1} \in \ell^0(\mathbb{Q})$ is said to be *Cauchy* if for any rational number $\varepsilon > 0$, there exists an $N \in \mathbb{N}$ such that

$$|\xi_j - \xi_k| < \varepsilon, \qquad \forall j, k \geqslant N.$$

The set of all Cauchy sequences of rational numbers is denoted by $c(\mathbb{Q})$.

(iii) Two sequences $\{\xi_n\}_{n \geqslant 1}, \{\eta_n\}_{n \geqslant 1} \in \ell^0(\mathbb{Q})$ are said to be *equivalent*, denoted by $\{\xi_n\}_{n \geqslant 1} \sim \{\eta_n\}_{n \geqslant 1}$, if for any rational number $\varepsilon > 0$, there exists an $N \in \mathbb{N}$ such that

$$|\xi_n - \eta_n| < \varepsilon, \qquad \forall n \geqslant N.$$

The following result is basic for the sequences of rational numbers, whose proof is left to the readers.

Proposition 3.3. (i) *Every Cauchy sequence is bounded.*

(ii) *If* $\{\xi_n\}_{n\geqslant 1}, \{\eta_n\}_{n\geqslant 1} \in \ell^0(\mathbb{Q})$ *are equivalent, then* $\{\xi_n\}_{n\geqslant 1} \in c(\mathbb{Q})$ *if and only if* $\{\eta_n\}_{n\geqslant 1} \in c(\mathbb{Q})$.

(iii) *Let* $\{\xi_n\}_{n\geqslant 1}, \{\eta_n\}_{n\geqslant 1} \in c(\mathbb{Q})$. *Then*
$$\{\xi_n + \eta_n\}_{n\geqslant 1},\ \{\xi_n \cdot \eta_n\}_{n\geqslant 1} \in c(\mathbb{Q}).$$

(iv) *If* $\{\xi_n\}_{n\geqslant 1} \in c(\mathbb{Q})$ *is bounded away from zero, i.e., there exists a rational number* $\delta > 0$ *such that*
$$|\xi_n| \geqslant \delta, \qquad \forall n \geqslant 1,$$
then $\{\xi_n^{-1}\}_{n\geqslant 1} \in c(\mathbb{Q})$.

(v) *If* $\{\xi_n\}_{n\geqslant 1}, \{\eta_n\}_{n\geqslant 1}, \{\bar{\xi}_n\}_{n\geqslant 1}, \{\bar{\eta}_n\}_{n\geqslant 1} \in \ell^\infty(\mathbb{Q})$ *such that*
$$\{\xi_n\}_{n\geqslant 1} \sim \{\bar{\xi}_n\}_{n\geqslant 1}, \quad \{\eta_n\}_{n\geqslant 1} \sim \{\bar{\eta}_n\}_{n\geqslant 1}.$$
Then
$$\{\xi_n + \eta_n\}_{n\geqslant 1} \sim \{\bar{\xi}_n + \bar{\eta}_n\}_{n\geqslant 1},$$
$$\{\xi_n \cdot \eta_n\}_{n\geqslant 1} \sim \{\bar{\xi}_n \cdot \bar{\eta}_n\}_{n\geqslant 1},$$
and if in addition, $\{\xi_n\}_{n\geqslant 1}$ *is bounded away from zero, then* $\{\bar{\xi}_n\}_{n\geqslant 1}$ *must also be bounded away from zero and*
$$\{\xi_n^{-1}\}_{n\geqslant 1} \sim \{\bar{\xi}_n^{-1}\}_{n\geqslant 1}.$$

(vi) *The relation "$\sim$" defined on* $c(\mathbb{Q})$ *is an equivalent relation, i.e., it possesses the reflexivity, symmetry and transitivity.*

3.2 A construction of real numbers

We now introduce the following.

Definition 3.4. (i) For any $\{\xi_n\}_{n\geqslant 1} \in c(\mathbb{Q})$, define the corresponding equivalent class by
$$\lim \xi_n = \Big\{ \{\eta_n\}_{n\geqslant 1} \in c(\mathbb{Q}) \mid \{\eta_n\}_{n\geqslant 1} \sim \{\xi_n\}_{n\geqslant 1} \Big\}.$$
Let
$$\mathbb{R} = \Big\{ \lim \xi_n \mid \{\xi_n\}_{\geqslant 1} \in c(\mathbb{Q}) \Big\}.$$
Any element in $\mathbb{R}$ is called a *real number*.

(ii) For any $r \in \mathbb{Q}$, define $\xi_n = r$ for all $n \geqslant 1$. Then $\{\xi_n\}_{n\geqslant 1} \in c(\mathbb{Q})$, which defines a real number. Thus, in such a sense,
$$\mathbb{Q} \subset \mathbb{R} \tag{3.1}$$
Any element in $\mathbb{R} \setminus \mathbb{Q}$ is called an *irrational number*.

It is clearly that

$$\{\xi_n\}_{n\geqslant 1} \sim \{\eta_n\}_{n\geqslant 1} \iff \lim \xi_n = \lim \eta_n.$$

Hence, we may identify $\lim \xi_n$ with any $\{\eta_n\}_{n\geqslant 1}$ which is equivalent to $\{\xi_n\}_{n\geqslant 1}$. In particular, $\lim \xi_n$ is identified with the sequence $\{\xi_n\}_{n\geqslant 1}$. We have the following basic result of reflexivity, symmetry, and transitivity for the real numbers with respect to the equality.

Proposition 3.5. *Let* $\{\xi_n\}_{n\geqslant 1}, \{\eta_n\}_{n\geqslant 1}, \{\zeta_n\}_{n\geqslant 1} \in c(\mathbb{Q})$ *and denote*

$$x = \lim \xi_n, \quad y = \lim \eta_n, \quad z = \lim \zeta_n.$$

Then

$$x = x;$$
$$x = y \quad \Rightarrow \quad y = x;$$
$$x = y, \; y = z \quad \Rightarrow \quad x = z.$$

Proof. Let us prove the transitivity. Let

$$x = \lim \xi_n, \quad y = \lim \eta_n, \quad z = \lim \zeta_n.$$

By $x = y$ and $y = z$, we know that

$$\{\xi_n\}_{n\geqslant 1} \sim \{\eta_n\}_{n\geqslant 1}, \quad \{\eta_n\}_{n\geqslant 1} \sim \{\zeta_n\}_{n\geqslant 1}.$$

Thus, $\{\xi_n\}_{n\geqslant 1} \sim \{\zeta_n\}_{n\geqslant 1}$, which means $x = z$. $\qquad\square$

The above actually shows that the real numbers are well-defined. We now introduce algebraic operations on $\mathbb{R}$.

Definition 3.6. For any $x, y \in \mathbb{R}$ with

$$x = \lim \xi_n, \quad y = \lim \eta_n,$$

for some $\{\xi_n\}_{n\geqslant 1}, \{\eta_n\}_{n\geqslant 1} \in c(\mathbb{Q})$, define addition, multiplication and negation as follows:

$$x + y = \lim(\xi_n + \eta_n), \quad x \cdot y = \lim(\xi_n \cdot \eta_n), \quad -x = \lim(-\xi_n). \tag{3.2}$$

In addition, if $\{\xi_n\}_{n\geqslant 1}$ is bounded away from zero, we say that x is non-zero, denoted by $x \neq 0$, and define the *reciprocal* of x by

$$x^{-1} = \lim \xi_n^{-1}. \tag{3.3}$$

According to the properties of the Cauchy sequence, the addition, multiplication, negation, and reciprocal of real numbers are all well-defined. Then we define *subtraction* and *division* by

$$x - y = x + (-y), \qquad x/y = xy^{-1}, \quad (y \neq 0).$$

We have results for real numbers similar to Proposition 2.14 for rational numbers. Since the statement is almost the same, except that $\mathbb{Q}$ is now replaced by $\mathbb{R}$, we shorten the statement below.

Proposition 3.7. *All the algebraic laws hold for real numbers.*

Definition 3.8. (i) A sequence $\{\xi_n\}_{n \geqslant 1} \in \ell^0(\mathbb{Q})$ is said to be *positively bounded away from zero* if there exists a positive rational number $\delta > 0$ and an $N \in \mathbb{N}$ such that

$$\xi_n \geqslant \delta, \qquad \forall n \geqslant N.$$

(ii) A real number x is *positive*, denoted by $x > 0$ if there exists a sequence $\{\xi_n\}_{n \geqslant 1} \in c(\mathbb{Q})$, positively bounded away from zero such that

$$x = \lim \xi_n.$$

If $x > 0$ or $x = 0$, we say that x is *non-negative*, denoted by $x \geqslant 0$. If $-x > 0$, we say x is *negative*, denoted by $x < 0$, and if $-x \geqslant 0$, we say that x is *non-positive*, denoted by $x \leqslant 0$.

(iii) For any $x \in \mathbb{R}$, define its *absolute value* $|x|$ by the following:

$$|x| = \begin{cases} x, & x \geqslant 0, \\ -x, & x < 0. \end{cases}$$

(iv) For any $x, y \in \mathbb{R}$, define $x > y$ (or $y < x$) if $x - y > 0$, and $x \geqslant y$ (or $y \leqslant x$) if $x > y$ or $x = y$.

(v) For any $x, y \in \mathbb{R}$, define

$$\max(x, y) = \begin{cases} x, & x \geqslant y, \\ y, & x < y, \end{cases} \qquad \min(x, y) = \begin{cases} y, & x \geqslant y, \\ x, & x < y. \end{cases}$$

It is interesting that the following hold:

$$\max(x, y) = \frac{x + y + |x - y|}{2},$$
$$\min(x, y) = \frac{x + y - |x - y|}{2}, \qquad \forall x, y \in \mathbb{R}.$$

Thus, both maps $(x, y) \mapsto \max(x, y)$ and $(x, y) \mapsto \min(x, y)$ can be written in terms of the absolute value.

Lemma 3.9. *Let* $\{\xi_n\}_{n \geq 1}, \{\eta_n\}_{n \geq 1} \in c(\mathbb{Q})$ *such that for some* $N \in \mathbb{N}$,

$$\xi_n \geq \eta_n, \qquad \forall n \geq N.$$

Then

$$\lim \xi_n \geq \lim \eta_n.$$

In particular, if

$$\xi_n \geq 0, \qquad \forall n \geq N,$$

then

$$\lim \xi_n \geq 0.$$

Proof. We prove the second conclusion. The first can be obtained from considering $\xi_n - \eta_n$.

Suppose $x = \lim \xi_n < 0$. Then by definition, there exists a positive rational number δ and a sequence $\{\zeta_n\}_{n \geq 1}$ such that

$$x = \lim \zeta_n, \qquad \zeta_n \leq -\delta, \qquad \forall n \geq 1.$$

Since $\{\xi_n\}_{n \geq 1} \sim \{\zeta_n\}_{n \geq 1}$, for $\varepsilon = \delta/2$, there exists an $\bar{N} \in \mathbb{N}$ such that

$$|\xi_n - \zeta_n| < \varepsilon = \delta/2, \qquad \forall n \geq \bar{N}.$$

This implies

$$\xi_n < \zeta_n + \varepsilon \leq -\delta + \delta/2 = -\delta/2 < 0, \qquad \forall n \geq \bar{N},$$

a contradiction. $\qquad \square$

Note that $\{\xi_n\}_{n \geq 1} \in c(\mathbb{Q})$ with $\xi_n > 0$, for all $n \geq 1$ is not enough to guarantee that $x = \lim \xi_n$ is positive. For example, for $\xi_n = n^{-1}$, $n \geq 1$, one has $\lim \xi_n = 0$.

The following is comparable with Proposition 2.16 (which is for rational numbers).

Proposition 3.10. *Let* $x, y, z \in \mathbb{R}$. *Then*

(i) (Addition preserves order) $x > y$ *if and only if* $x + z > y + z$.

(ii) (Positive multiplication preserves order) *Let* $z > 0$. *Then* $x > y$ *if and only if* $x \cdot z > y \cdot z$.

(iii) (Negation reverses order) $x > y$ if and only if $-x < -y$.

(iv) (Order is transitive) If $x > y$ and $y > z$, then $x > z$.

(v) (Trichotomy of reals) *Exactly one of the three statements holds:* $x < y$, $x = y$, and $x > y$.

(vi) If $x > 0$, then $x^{-1} > 0$. If $x > y > 0$, then $x^{-1} < y^{-1}$.

Proof. We only prove (vi). The proof for the other parts is left to the readers. Since $x > 0$, there exists a sequence $\{\xi_n\}_{n \geqslant 1} \in c(\mathbb{Q})$ such that $x = \lim \xi_n$ and for some rational numbers $M, \delta > 0$,

$$M \geqslant \xi_n \geqslant \delta, \qquad \forall n \geqslant 1.$$

Then we know that $\{\xi_n^{-1}\}_{n \geqslant 1} \in c(\mathbb{Q})$ and

$$M^{-1} \leqslant \xi_n^{-1} \leqslant \delta^{-1}, \qquad n \geqslant 1.$$

This implies that

$$x^{-1} = \lim \xi_n^{-1} > 0.$$

Now, let $x > y > 0$. Then we have sequences $\{\xi_n\}_{n \geqslant 1}, \{\eta_n\}_{n \geqslant 1} \in c(\mathbb{Q})$ such that for some rational numbers $M, \delta > 0$, one has

$$M \geqslant \xi_n, \eta_n \geqslant \delta, \qquad \xi_n - \eta_n \geqslant \delta, \qquad \forall n \geqslant 1.$$

Hence, for some $N \in \mathbb{N}$,

$$\xi_n > \eta_n + \delta/2 \geqslant \delta, \qquad \forall n \geqslant N.$$

Consequently,

$$M^{-1} \leqslant \xi_n^{-1} < (\eta_n + \delta/2)^{-1} < \eta_n^{-1}, \qquad \forall n \geqslant N.$$

Hence,

$$M^{-1} \leqslant x^{-1} \leqslant (y + \delta/2)^{-1} < y^{-1}.$$

This proves (vi). $\square$

We now extend the exponentiation from $\mathbb{Q}$ to $\mathbb{R}$.

Definition 3.11. For any $x \in \mathbb{R} \setminus \{0\}$, define

$$x^0 = 1, \qquad x^{n+1} = x^n \cdot x, \qquad \forall n \geqslant 0,$$

and

$$x^{-n} = (x^n)^{-1} \equiv 1/x^n.$$

The above defines x^a for any $x \in \mathbb{R} \setminus \{0\}$ and any $a \in \mathbb{Z}$. The following lemma is useful below.

Lemma 3.12. *Let* $x = \lim \xi_n \in \mathbb{R} \setminus \{0\}$ *with* $\{\xi_n\}_{n \geqslant 1} \in c(\mathbb{Q})$*, which is away from zero. Then for any* $a \in \mathbb{Z}$*,* $\{\xi_n^a\}_{n \geqslant 1} \in c(\mathbb{Q})$ *and*

$$x^a = \lim \xi_n^a.$$

Proof. First, we let $a = m \in \mathbb{N}$ and use induction on m. For $m = 0$, it is trivial. Suppose the conclusion holds for m. Then for $m + 1$, we

$$x^{m+1} = x^m \cdot x = \lim(\xi_n^m) \cdot \lim(\xi_n) = \lim(\xi_n^m \cdot \xi_n) = \lim(\xi_n^{m+1}).$$

Now, for $a = -m$ with $m \in \mathbb{Z}$, we have

$$x^a = x^{-m} = (x^m)^{-1} = \lim(\xi_n^m)^{-1} = \lim(\xi_n^{-m}) = \lim \xi_n^a.$$

This completes the proof. □

We now present the following result which is comparable with Proposition 2.20.

Proposition 3.13. *Let* $x, y \in \mathbb{R} \setminus \{0\}$ *and* $a, b \in \mathbb{Z}$*. Then the following hold:*

(i) $x^a \cdot x^b = x^{a+b}$, $(x^a)^b = x^{ab}$, $(x \cdot y)^a = x^a \cdot y^a$, $|x^a| = |x|^a$.

(ii) *If* $a \neq 0$*, then* $x^a = y^a$ *if and only if* $x = y$.

(iii) *For* $n \in \mathbb{N}$,

$$x > y > 0 \quad \Rightarrow \quad x^n > y^n > 0, \quad y^{-n} > x^{-n} > 0.$$

Proof. Let us prove the first equality in (i) to get the flavor. Other proofs are left to the readers. Let $\{\xi_n\}_{n \geqslant 1} \in c(\mathbb{Q})$ such that

$$x = \lim \xi_n.$$

Then by Proposition 2.20, we know that

$$\xi_n^a \cdot \xi_n^b = \xi_n^{a+b}.$$

Thus, by Lemma 3.12,

$$x^a \cdot x^b = \lim(\xi_n^a) \cdot \lim(\xi_n^b) = \lim(\xi_n^a \cdot \xi_n^b) = \lim(\xi_n^{a+b}) = x^{a+b},$$

proving our conclusion. □

3.3 *Further properties of real numbers*

We now look at some further properties of the set $\mathbb{R}$. Recall that there is no rational number ξ such that $\xi^2 = 2$. Therefore, we would like to see if there exists a real number x such that $x^2 = 2$. Before going further, we need to make some preparations.

Proposition 3.14. (i) (Bounding of reals by integers) *For any $x \in \mathbb{R}$, there exists a unique integer, denoted by $[x]$ such that*

$$[x] \leqslant x < [x] + 1. \tag{3.4}$$

(ii) (Archimedean property) *For any real numbers $x, \varepsilon > 0$, there exists a natural number N such that $N\varepsilon > x$.*

(iii) (Density of rational numbers) *For any real numbers $x < y$, there exists a rational number q such that $x < q < y$.*

Proof. (i) Let $x > 0$. There exists a sequence $\{\xi_n\}_{n \geqslant 1} \in c(\mathbb{Q})$ such that

$$x = \lim \xi_n.$$

Since Cauchy sequence is bounded, we can find rational numbers q, r and a natural number N such that

$$0 < q \leqslant \xi_n \leqslant r, \qquad \forall n \geqslant N.$$

Then, by Proposition 2.21, we have

$$[q] \leqslant q \leqslant \xi_n \leqslant r < [r] + 1.$$

Hence,

$$[q] \leqslant x \leqslant [r] + 1.$$

Now, comparing $[q] + 1$ with x, there are two possibilities:

$$[q] + 1 > x, \qquad \text{or} \qquad [q] + 1 \leqslant x.$$

If the former holds, then we take $[x] = [q]$, if the latter holds, we further compare $[q] + 2$ with x. Since $[r] \in \overline{\mathbb{N}}$, finitely many comparisons will give us an integer p such that

$$p \leqslant x < p + 1.$$

Clearly, such a p is unique. Hence, we call $p = [x]$.

The case $x < 0$ can be treated similarly.

(ii) Since $x/\varepsilon > 0$ is a given a real number, there exists a natural number N such that

$$x/\varepsilon < N.$$

Then

$$x < N\varepsilon.$$

(iii) Since $y - x > 0$, there exist sequences $\{\xi_n\}_{n\geqslant 1}, \{\eta_n\}_{n\geqslant 1} \in c(\mathbb{Q})$ such that

$$x = \lim \xi_n, \qquad y = \lim \eta_n,$$

and for some positive rational number $\delta > 0$,

$$\eta_n - \xi_n \geqslant \delta, \qquad \forall n \geqslant 1.$$

Then

$$\xi_n \leqslant \xi_n + \delta/4 < \eta_n - \delta/4 \leqslant \eta_n.$$

Also, since both $\{\xi_n\}_{n\geqslant 1}$ and $\{\eta_n\}_{n\geqslant 1}$ are Cauchy, we may assume that

$$|\xi_n - \xi_N|, |\eta_n - \eta_N| < \delta/4, \qquad \forall n \geqslant N.$$

Hence,

$$\xi_n < \xi_N + \delta/4 \leqslant \eta_N - \delta + \delta/4 < \eta_n + \delta/4 - \delta + \delta/4 = \eta_n - \delta/2 < \eta_n.$$

Therefore,

$$x = \lim \xi_n \leqslant \xi_N + \delta/4 \leqslant \eta_N - 3\delta/4 \leqslant \lim \eta_n = y.$$

Consequently, we may take

$$q = \frac{1}{2}\left(\eta_N - \frac{3\delta}{4} - \left(\xi_N + \frac{\delta}{4}\right)\right) = \frac{\eta_N - \xi_N}{2} - \frac{\delta}{4}.$$

This proves (iii). $\square$

Definition 3.15. (i) Let $E \subseteq \mathbb{R}$. An $M \in \mathbb{R}$ is called an *upper bound* for E if

$$x \leqslant M, \qquad \forall x \in E.$$

When such an M exists, E is said to be *bounded from above*. If no such an $M \in \mathbb{R}$ exists, E is said to be *unbounded from above*.

(ii) Let $E \subseteq \mathbb{R}$. A real number M is called a *least upper bound* of E if M is an upper bound of E and for any upper bound M' of E, it holds:

$$M \leqslant M'.$$

Example 3.16. (i) Let $\mathbb{R}_+ = \{x \in \mathbb{R} \mid x \geqslant 0\}$. Then $\mathbb{R}_+$ does not have an upper bound. Thus, $\mathbb{R}_+$ is unbounded from above.

(ii) Let $[0,1] \triangleq \{x \in \mathbb{R} \mid 0 \leqslant x \leqslant 1\}$. Then $[0,1]$ has an upper bound, and 1 is a least upper bound of $[0,1]$.

(iii) Any $M \in \mathbb{R}$ is an upper bound of empty set $\varnothing$.

The following result is very important.

Theorem 3.17. *Let $E \subseteq \mathbb{R}$.*

(i) *E can have at most one least upper bound.*

(ii) *If E has an upper bound, then it has a least upper bound.*

Proof. (i) Let M_1 and M_2 be two least upper bounds. Since they are upper bounds of E, we must have

$$M_1 \leqslant M_2, \qquad M_2 \leqslant M_1.$$

Hence, $M_1 = M_2$.

(ii) Without loss of generality, we may assume that there exists an $x_0 \in E$ such that $x_0 > 0$. In fact, take any $x_0 \in E$ and define

$$E_0 = \{x + |x_0| + 1 \mid x \in E\}.$$

If we can find a least upper bound s_0 of E_0, then E has a least upper bound s given by the following:

$$s = s_0 - |x_0| - 1.$$

We now fix an $x_0 \in E$ with $x_0 > 0$. Let $M \geqslant x_0 > 0$ be an upper bound of E. For any $n \in \mathbb{N}$ such that $n > \bar{n} = [\frac{1}{x_0}] + 1$, one has

$$\frac{1}{x_0} \leqslant \left[\frac{1}{x_0}\right] + 1 < n,$$

which means that $x_0 > 1/n$. Thus, $1/n$ is not an upper bound of E. Now, by Proposition 3.14, (ii) (the Archimedean property), there exists a $K \in \mathbb{N}$, such that

$$K/n \geqslant M. \tag{3.5}$$

We claim that there exists an $m_n \in \mathbb{N}$ such that

$$0 < m_n \leqslant K,$$

with m_n/n being an upper bound of E and $(m_n - 1)/n$ not being upper bound of E. To show this, we begin with K. By (3.5), we know that K/n is an upper bound of E. If $(K-1)/n$ is not an upper bound of E, then we let $m_n = K$. Suppose $(K-1)/n$ is still an upper bound of E. Then we check if $(K-2)/n$ is an upper bound of E. If not, we let $m_n = K - 1$. Otherwise, we continue the procedure to check if $(K-3)/n$ is an upper bound of E, and so on. Since $1/n$ is not an upper bound of E, the above procedure will stop at some m_n with $0 < m_n \leqslant K$.

Now, for the sequence $\{m_n\}_{n \geqslant \bar{n}}$, for any $n, n' \geqslant N$ (with $N \geqslant \bar{n}$ large), we have

$$\frac{m_n}{n} > \frac{(m_{n'} - 1)}{n'}, \qquad \frac{m_{n'}}{n'} > \frac{m_n - 1}{n},$$

since the left-hand sides are upper bounds of E and the right-hand sides are not. Thus,

$$\frac{m_n}{n} - \frac{m_{n'}}{n'} > -\frac{1}{n'} \geqslant -\frac{1}{N},$$

and

$$\frac{m_n}{n} - \frac{m_{n'}}{n'} < \frac{1}{n} \leqslant \frac{1}{N}.$$

Hence, $\{m_n/n\}_{n \geqslant \bar{n}}$ is Cauchy. Let

$$s = \lim \frac{m_n}{n} = \lim \frac{m_n - 1}{n}.$$

We claim that s is the least upper bound of E. In fact, since for any $x \in E$,

$$x \leqslant \frac{m_n}{n}, \qquad \forall n \geqslant \bar{n}.$$

This leads to

$$x \leqslant \lim \frac{m_n}{n} = s.$$

On the other hand, since $\frac{m_n-1}{n}$ is not an upper bound of E, for any upper bound M' of E, we have

$$\frac{m_n - 1}{n} \leqslant M', \qquad \forall n \geqslant \bar{n}.$$

Hence,

$$s = \lim \frac{m_n - 1}{n} \leqslant M'.$$

This shows that s is the least upper bound of E. $\qquad \square$

Because of the above result, we hereafter denote the least upper bound of E by $\sup(E)$, call it the *supremum* of E. If E has no upper bound, we let

$$\sup(E) = +\infty,$$

and in the case $E = \varnothing$, we define

$$\sup(E) = -\infty,$$

since any $M \in \mathbb{R}$ is an upper bound of $\varnothing$.

Now, we are ready to prove the following proposition.

Proposition 3.18. *There exists a positive $x \in \mathbb{R}$ such that $x^2 = 2$.*

Proof. Let

$$E = \{y \in \mathbb{R} \mid y \geqslant 0, \ y^2 < 2\}.$$

Since $0 \in E$, $E \neq \varnothing$. Also, if $y > 2$, one has

$$y^2 > 4 > 2.$$

Hence, 2 is an upper bound of E. Now, let

$$x = \sup(E) \in \mathbb{R}.$$

We claim that $x^2 = 2$. In fact, suppose $x^2 < 2$. Let $0 < \varepsilon < 1$ be small. Then (note that 2 is an upper bound of E, $x \leqslant 2$)

$$(x + \varepsilon)^2 = x^2 + 2\varepsilon x + \varepsilon^2 \leqslant x^2 + 4\varepsilon + \varepsilon \leqslant x^2 + 5\varepsilon.$$

Thus, by choosing $\varepsilon > 0$ small, we have

$$(x + \varepsilon)^2 < 2.$$

Hence, $x + \varepsilon \in E$, contradicting the definition of x.

Now, suppose $x^2 > 2$. Then again let $0 < \varepsilon < 1$ be small. We have

$$(x - \varepsilon)^2 = x^2 - 2\varepsilon x + \varepsilon^2 \geqslant x^2 - 2\varepsilon x \geqslant x^2 - 4\varepsilon,$$

since $x \leqslant 2$. Hence, by choosing $\varepsilon > 0$ small, we will have

$$(x - \varepsilon)^2 > 2.$$

But then, $x - \varepsilon$ is an upper bound of E (since if there exists an $y \in E$ such that $x - \varepsilon \leqslant y$, then $(x - \varepsilon)^2 < y^2 < 2$, a contradiction). This contradicts the definition of x again. Hence, $x^2 = 2$. $\qquad\square$

The following result is useful.

Theorem 3.19. *Let $E \subseteq \mathbb{R}$ have an upper bound, and $b \in \mathbb{R}$. Then $b = \sup(E)$ if and only if*

$$x \leqslant b, \qquad \forall x \in E, \tag{3.6}$$

and for any $\varepsilon > 0$, there exists an $x \in E$ such that

$$x > b - \varepsilon. \tag{3.7}$$

Proof. Sufficiency. Let $M = \sup(E)$. By our assumption, b is an upper bound of E, we must have

$$M \leqslant b.$$

Next, for any $\varepsilon > 0$, since there exists an $x \in E$ such that (3.7) holds, which means that $b - \varepsilon$ is not an upper bound of E, whereas M is an upper bound of E. Hence, one must have

$$b - \varepsilon < M \leqslant b.$$

Since $\varepsilon > 0$ is arbitrary, we obtain $b = M$.

Necessity. If $b = \sup(E)$, then it is an upper bound of E. Therefore, (3.6) holds. Also, since b is the least upper bound which means that for any $\varepsilon > 0$, $b - \varepsilon$ is not an upper bound of E. This implies that there exists an $x \in E$ such that (3.7) holds.   □

Symmetrically, we may define *lower bound* of a set $E \subseteq \mathbb{R}$ and the *greatest lower bound* for E. We have the following result.

Theorem 3.20. *Let $E \subseteq \mathbb{R}$.*

(i) *E can have at most one greatest lower bound.*

(ii) *If E has a lower bound, then it has a greatest lower bound.*

The proof is very similar to that of Theorem 3.17. One may also do the following. Let

$$-E = \{-x \in \mid x \in E\}.$$

Then $-E$ has an upper bound if E has a lower bound. Therefore, there exists a unique least upper bound $\sup(-E)$. Then we let

$$K = -\sup(-E).$$

One can check that K is the unique greatest lower bound of E. Hereafter, we denote the greatest lower bound of E by $\inf(E)$, and called the *infimum* of E. Similar to Theorem 3.19, we also have the following result.

Theorem 3.21. *Let $E \subseteq \mathbb{R}$ have a lower bound, and $a \in \mathbb{R}$. Then $a = \inf(E)$ if and only if*

$$x \geqslant a, \qquad \forall x \in E,$$

and for any $\varepsilon > 0$, there exists an $x \in E$ such that

$$x < a + \varepsilon.$$

Proposition 3.18, together with the results concerning the supremum leads to some more general result. We now establish such kind of result.

Lemma 3.22. *Let $x \in \mathbb{R}$, $x > 0$ and let $n \in \mathbb{N}$, $n \geqslant 2$. Then the set*
$$E_n(x) = \{y \in \mathbb{R} \mid y \geqslant 0, \ y^n \leqslant x\}$$
is non-empty and bounded above.

Proof. Since $0 \in E_n(x)$, we have the non-emptiness of $E_n(x)$. Further, we claim that E has an upper bound. We look at two cases.

Case 1. $x \leqslant 1$. Then we must have
$$y \leqslant 1, \qquad \forall y \in E_n(x).$$
In fact, if the above is not the case, i.e., there exists a $y \in E_n(x)$ such that $y > 1$, then $x \geqslant y^n > 1$, a contradiction.

Case 2. $x > 1$. Then we must have
$$y \leqslant x, \qquad \forall y \in E_n(x).$$
In fact, if the above is not the case, i.e., there exists a $y \in E_n(x)$ such that
$$y > x > 1.$$
Then $y^{n-1} > 1$. Thus
$$y^n = y^{n-1}y > y > x,$$
which contradicts $y \in E_n(x)$. □

Combining the above result with Theorem 3.17, we see that the following definition makes sense.

Definition 3.23. *Let $x \in \mathbb{R}$, $x > 0$, and $n \in \mathbb{N}$, $n \geqslant 2$. Define*
$$x^{\frac{1}{n}} = \sup\{y \in \mathbb{R} \mid y \geqslant 0, \ y^n \leqslant x\},$$
which is called the n-th root of x. Also, we define
$$x^{-\frac{1}{n}} = (x^{\frac{1}{n}})^{-1} = \frac{1}{x^{\frac{1}{n}}}.$$

The following result is concerned with the n-th root operation.

Proposition 3.24. *Let $x, y \in \mathbb{R}$, $x, y > 0$, and $n, m \in \mathbb{N}$, $n, m \geqslant 1$. Then the following hold:*

(i) $y = x^{\frac{1}{n}}$ *if and only if* $y^n = x$. *Thus,* $(x^{\frac{1}{n}})^n = x$.

(ii) $x > y$ *if and only if* $x^{\frac{1}{n}} > y^{\frac{1}{n}}$. *In particular,* $x^{\frac{1}{n}} > 0$.

(iii) $(x \cdot y)^{\frac{1}{n}} = x^{\frac{1}{n}} y^{\frac{1}{n}}$, $(x^{\frac{1}{n}})^{\frac{1}{m}} = x^{\frac{1}{nm}}$.

(iv) *For $n \geqslant m$, if $x > 1$, then $x^{\frac{1}{n}} < x^{\frac{1}{m}}$; if $x < 1$, then $x^{\frac{1}{n}} > x^{\frac{1}{m}}$; and if $x = 1$, then $x^{\frac{1}{k}} = 1$ for all $k \in \mathbb{N}$.*

Proof. (i) Suppose $y = x^{\frac{1}{n}}$. We show that $y^n = x$. If this is not the case, then either $y^n < x$ or $y^n > x$. In the first case, for $\varepsilon \in (0, 1)$ small enough, we have

$$(y + \varepsilon)^n = y^n + \varepsilon \sum_{k=1}^{n} \frac{n!}{k!(n-k)!} \varepsilon^{k-1} y^{n-k} < x,$$

provided $\varepsilon > 0$ small enough, where $n! = n \cdot (n-1) \cdots 2 \cdot 1$ is called the *factorial* of n. Hence,

$$y + \varepsilon \in \left\{ z \geqslant 0 \mid z \geqslant 0, \ z^n \leqslant x \right\} \equiv E_n(x).$$

Consequently,

$$y + \varepsilon \leqslant \sup \left(E_n(x) \right) = x^{\frac{1}{n}} = y,$$

a contradiction. Now, in the second case, for $\varepsilon \in (0, 1)$ small enough,

$$(y - \varepsilon)^n = y^n + \varepsilon \sum_{k=1}^{n} (-1)^k \frac{n!}{k!(n-k)!} \varepsilon^{k-1} y^{n-k} > x.$$

Hence, $y - \varepsilon \notin E_n(x)$. This leads to

$$y - \varepsilon \geqslant \sup \left(E_n(x) \right) = x^{\frac{1}{n}} = y,$$

a contradiction again. Hence, $y^n = x$.

Conversely, if $y^n = x$, then $y \in E_n(x)$ which implies

$$y \leqslant \sup \left(E_n(x) \right) = x^{\frac{1}{n}}.$$

On the other hand, for any $z \in E_n(x)$,

$$z^n \leqslant x = y^n.$$

By Proposition 3.13, (iii), we must have $z \leqslant y$. This means that y is an upper bound of $E_n(x)$. Hence,

$$x^{\frac{1}{n}} = \sup \left(E_n(x) \right) \leqslant y.$$

Hence, $y = \sup E_n(x) = x^{\frac{1}{n}}$.

(ii) First, let $x^{\frac{1}{n}} > y^{\frac{1}{n}}$. Then by Proposition 3.13, (iii), we have

$$x = (x^{\frac{1}{n}})^n > (y^{\frac{1}{n}})^n = y.$$

Conversely, let $x > y$. Thus,

$$E_n(y) \equiv \left\{ z \in \mathbb{R} \mid z \geqslant 0, \ z^n \leqslant y \right\} \subseteq E_n(x).$$

This implies that $x^{\frac{1}{n}}$ is an upper bound of $E_n(y)$. Hence,

$$y^{\frac{1}{n}} \leqslant x^{\frac{1}{n}}.$$

We claim that the strict inequality holds in the above. If not, then $y^{\frac{1}{n}} = x^{\frac{1}{n}}$, which implies that $y = x$, a contradiction.

The proofs of (iii) and (iv) are left to the readers. $\qquad\square$

With the above results, we now introduce the following definition.

Definition 3.25. Let $x \in \mathbb{R}$, $x > 0$ and let $\xi \in \mathbb{Q}$ with $\xi = a/b$ for some $a, b \in \mathbb{Z}$, $b \neq 0$. Define

$$x^{\xi} = (x^{\frac{1}{b}})^a.$$

The following shows that the above defined rational exponent of a real number is well-defined.

Lemma 3.26. *Let* $a, a', b, b' \in \mathbb{Z}$, $b, b' > 0$ *such that*

$$\frac{a}{b} = \frac{a'}{b'}.$$

Then for any $x \in \mathbb{R}$, $x > 0$,

$$(x^{\frac{1}{b}})^a = (x^{\frac{1}{b'}})^{a'}.$$

Proof. Consider three cases.

Case 1. $a = 0$. Then $a' = 0$. Therefore,

$$(x^{\frac{1}{b}})^a = x^0 = (x^{\frac{1}{b'}})^{a'}.$$

Case 2. $a > 0$. Then $a' > 0$. From

$$ab' = a'b,$$

we have

$$y \equiv x^{\frac{1}{ab'}} = (x^{\frac{1}{b'}})^{\frac{1}{a}}, \qquad y \equiv x^{\frac{1}{a'b}} = (x^{\frac{1}{b}})^{\frac{1}{a'}}.$$

Hence, by Proposition 3.24, (i),

$$y^a = x^{\frac{1}{b'}}, \qquad y^{a'} = x^{\frac{1}{b}}.$$

Therefore,

$$(x^{\frac{1}{b'}})^{a'} = (y^a)^{a'} = y^{aa'} = (y^{a'})^a = (x^{\frac{1}{b}})^a.$$

Case 3. $a < 0$. Then $(-a)/b = (-a')/b'$. Since $-a, -a' > 0$, we obtain

$$(x^{\frac{1}{b}})^{-a} = (x^{\frac{1}{b'}})^{-a'}.$$

Taking reciprocal, we are done.   □

The following collects some results for the rational exponents of real numbers.

Proposition 3.27. *Let $x, y \in \mathbb{R}$ with $x, y > 0$ and let $\xi, \eta \in \mathbb{Q}$. Then*

(i) $x^\xi \cdot x^\eta = x^{\xi+\eta}$, $(x^\xi)^\eta = x^{\xi\eta}$, $x^{-\xi} = 1/x^\xi$.

(ii) *If $\xi > 0$, then $x > y$ if and only if $x^\xi > y^\xi$. In particular, $x^\xi > 0$.*

(iii) *If $x > 1$ then $x^\xi > x^\eta$ if and only if $\xi > \eta$. If $x < 1$, then $x^\xi > x^\eta$ if and only if $\xi < \eta$.*

3.4   Real exponentiation

We have defined x^ξ for any $x \in \mathbb{R}$, $x > 0$ and any $\xi \in \mathbb{Q}$. We now would like to define x^y for any $x, y \in \mathbb{R}$, $x > 0$. Let us begin with the following lemma.

Lemma 3.28. *Let $x, y \in \mathbb{R}$ with $x > 0$, and $y = \lim \eta_n$ with $\{\eta_n\}_{n \geqslant 1} \in c(\mathbb{Q})$. Then $\{x^{\eta_n}\}_{n \geqslant 1} \in c(\mathbb{R})$. Furthermore, if $\{\zeta_n\}_{n \geqslant 1} \sim \{\eta_n\}_{n \geqslant 1}$, then $\{x^{\zeta_n}\}_{n \geqslant 1} \in c(\mathbb{R})$ and*

$$\lim_{n \to \infty} x^{\zeta_n} = \lim_{n \to \infty} x^{\eta_n}.$$

Proof. Let $x > 1$. Since $\{\eta_n\}_{n \geqslant 1}$ is Cauchy, it is bounded by, say, $M > 0$:

$$|\eta_n| \leqslant M, \qquad \forall n \geqslant 1.$$

Also, since $\lim_{n \to \infty} x^{\frac{1}{n}} = 1$, for any $\varepsilon > 0$, there exists a $K \geqslant 1$ such that

$$|x^{\frac{1}{K}} - 1| < \varepsilon x^{-M},$$

and there exists an $N \in \mathbb{N}$ such that

$$|\eta_n - \eta_m| < \frac{1}{K}, \qquad \forall n, m \geqslant N.$$

Consequently, by assuming, $\eta_n \geqslant \eta_m$, we have

$$|x^{\eta_n} - x^{\eta_m}| = x^{\eta_m}|x^{\eta_n - \eta_m} - 1| \leqslant x^M |x^{\frac{1}{K}} - 1| < \varepsilon, \quad \forall m, n \geqslant N.$$

Hence, $\{x^{\eta_n}\}_{n \geqslant 1}$ is Cauchy, and it is convergent.

Next, suppose $\{\zeta_n\}_{n \geqslant 1} \sim \{\eta_n\}_{n \geqslant 1}$. Let $r_n = \eta_n - \zeta_n$. Then $r_n \to 0$. We claim that

$$\lim_{n \to \infty} x^{r_n} = 1. \tag{3.8}$$

If this is the case, then by $x^{\eta_n} = x^{\zeta_n} x^{r_n}$, we have

$$\lim_{n\to\infty} x^{\eta_n} = \lim_{n\to\infty} x^{\zeta_n} \lim_{n\to\infty} x^{r_n} = \lim_{n\to\infty} x^{\zeta_n}.$$

Now, we prove (3.8). To this end we find a sequence k_n of natural numbers such that

$$|r_n| \leqslant \frac{1}{k_n}, \qquad \forall n \geqslant 1.$$

Then (noting $x > 1$)

$$\left(\frac{1}{x}\right)^{\frac{1}{k_n}} \leqslant \left(\frac{1}{x}\right)^{|r_n|} = x^{-|r_n|} \leqslant 1 \leqslant x^{\frac{1}{k_n}}.$$

By Squeeze Theorem (Corollary 2.11 of Chapter 1), we have

$$\lim_{n\to\infty} x^{r_n} = 1.$$

This completes the proof. $\qquad\qquad\qquad\qquad\qquad\qquad\qquad\qquad\square$

Because of the above result, we now introduce the following real exponentiation of real numbers.

Definition 3.29. Let $x, y \in \mathbb{R}$ with $x > 0$. Let $y = \lim \eta_n$ with $\eta_n \in \mathbb{Q}$, $n \geqslant 1$. We define:

$$x^y = \lim_{n\to\infty} x^{\eta_n}. \tag{3.9}$$

The following gives some properties of real exponents of positive real numbers.

Theorem 3.30. *Let* $x, y > 0$ *and* $p, q \in \mathbb{R}$. *Then*

(i) $x^p \in \mathbb{R}$, $x^p > 0$.

(ii) $x^p \cdot x^q = x^{p+q}$, $(x^p)^q = x^{pq}$.

(iii) $x^{-p} = 1/x^p$.

(iv) *If* $p > 0$, *then* $x > y$ *if and only if* $x^p > y^p$.

(v) *If* $x > 1$, *then* $x^p > x^q$ *if and only if* $p > q$; *if* $x < 1$, *then* $x^p > x^q$ *if and only if* $p < q$.

Proof. We leave the proof of (i)–(iii) to the readers.

(iv) Since $p > 0$, we have $p_n \in \mathbb{Q}$, bounded away from 0 such that

$$\lim_{n\to\infty} p_n = p, \qquad \lim_{n\to\infty} \frac{1}{p_n} = \frac{1}{p}.$$

Consequently,

$$x^p = \lim_{n \to \infty} x^{p_n}, \quad y^p = \lim_{n \to \infty} y^{p_n},$$
$$x = \lim_{n \to \infty} x^{\frac{p}{p_n}}, \quad y = \lim_{n \to \infty} y^{\frac{p}{p_n}}.$$

It is known that

$$x > y, \quad \Longleftrightarrow \quad x^{p_n} > y^{p_n}, \quad \forall n \geqslant 1.$$

Hence,

$$x^p = \lim_{n \to \infty} x^{p_n} \geqslant \lim_{n \to \infty} y^{p_n} = y^p.$$

We claim that

$$x^p > y^p. \tag{3.10}$$

If not, i.e., $x^p = y^p$, then

$$x = x^{\frac{p}{p}} = \lim_{n \to \infty} (x^p)^{\frac{1}{p_n}} = \lim_{n \to \infty} (y^p)^{\frac{1}{p_n}} = y,$$

contradicts $x > y$. Hence, (3.10) holds. The other direction can be proved in the same way.

(v) Let $x > 1$. We only prove the case that $x^p > 1$ if and only if $p > 0$ (i.e., $q = 0$). If $p > 0$, we can find $p_n, \delta \in \mathbb{Q}$ such that

$$p = \lim_{n \to \infty} p_n, \quad p_n \geqslant \delta > 0, \quad \forall n \geqslant 1.$$

Then

$$x^p = \lim_{n \to \infty} x^{p_n} \geqslant x^\delta > 1.$$

Conversely, let $x^p > 1$. We claim that $p > 0$. First of all $p = 0$ is impossible. If $p < 0$, then we have $p_n, \delta \in \mathbb{Q}$ such that

$$p = \lim_{n \to \infty} p_n, \quad p_n \leqslant \delta < 0, \quad \forall n \geqslant 1.$$

Then

$$x^p = \lim_{n \to \infty} x^{p_n} \leqslant x^\delta < 1.$$

This proves the special case. The other cases can be proved similarly. $\qquad \square$

We recall that the horizontal straight line $\mathbb{L}$ has the property that $\mathbb{Q} \subset \mathbb{L}$ by marking every rational number on $\mathbb{L}$. Also, we have defined a natural order on $\mathbb{L}$: Any two points $x, y \in \mathbb{L}$ satisfying $x < y$ if and only if x is on the left of y. On the other hand, one has $\mathbb{Q} \subset \mathbb{R}$, by identifying any $r \in \mathbb{Q}$ with an equivalent class of Cauchy sequence for which a representative $\{\xi_n\}_{n \geqslant 1}$

is given by $\xi_n = r$ for all $n \geqslant 1$. A natural question is what is the relation between $\mathbb{R}$ and $\mathbb{L}$? To look at this, we pick any $x \in \mathbb{R} \backslash \mathbb{Q}$, let $\{\xi_n\}_{n \geqslant 1} \in c(\mathbb{Q})$ such that

$$x \equiv \lim \xi_n = \lim_{n \to \infty} \xi_n.$$

Now, we construct $\{\eta_n\} \in c(\mathbb{Q})$, which is increasing, and $\{\zeta_n\} \in c(\mathbb{Q})$, which is decreasing, such that both are equivalent to $\{\xi_n\}$. Since $\{\xi_n\}$ is Cauchy, for any $k \geqslant 1$, there exists an $N_k \in \mathbb{N}$ such that

$$|\xi_i - \xi_j| < \frac{1}{k(k+1)}, \qquad \forall i, j \geqslant N_k.$$

We may take $N_1 < N_2 < N_3 < \cdots$. Thus,

$$\xi_{N_k} - \frac{1}{k(k+1)} < \xi_{N_{k+1}} < \xi_{N_k} + \frac{1}{k(k+1)}.$$

This implies

$$\xi_{N_k} - \frac{1}{k} < \xi_{N_{k+1}} - \frac{1}{k+1} < \xi_{N_{k+1}} + \frac{1}{k+1} < \xi_{N_k} + \frac{1}{k}, \quad \forall k \geqslant 1.$$

Hence, by taking

$$\eta_k = \xi_{N_k} - \frac{1}{k}, \quad \zeta_k = \xi_{N_k} + \frac{1}{k}, \qquad k \geqslant 1,$$

we see that $\{\eta_k\}$ is increasing and $\{\zeta_k\}$ is decreasing. Further

$$\lim_{k \to \infty} \eta_k = \lim_{k \to \infty} \zeta_k = \lim_{n \to \infty} \xi_n = x.$$

Thus, it is easy to see that $\{\eta_n\} \sim \{\xi_n\} \sim \{\zeta_n\}$. Now, $\eta_k, \zeta_k \in \mathbb{Q}$ are well-marked on $\mathbb{L}$ and

$$\zeta_k - \eta_k = \frac{2}{k} \to 0.$$

Hence, there exists a unique point on $\mathbb{L}$, denoted by $\bar{x}$, such that

$$\eta_1 < \eta_2 < \cdots < \eta_k < \bar{x} < \zeta_k < \cdots < \zeta_2 < \zeta_1, \qquad \forall k \geqslant 1.$$

Then we naturally identify the point $\bar{x} \in \mathbb{L}$ with the real number $x \in \mathbb{R}$. This shows that $\mathbb{R} \subseteq \mathbb{L}$. On the other hand, any point on $\mathbb{L}$ must be a limit point of a rational sequence. Therefore, by the property that $\mathbb{R}$ is closed under limit process, we obtain that $\mathbb{L} \subseteq \mathbb{R}$. Hence, in such a way, we finally obtain

$$\mathbb{N} \subseteq \bar{\mathbb{N}} \subset \mathbb{Z} \subset \mathbb{Q} \subset \mathbb{R} = \mathbb{L}. \tag{3.11}$$

The equality $\mathbb{R} = \mathbb{L}$ explains the completeness of the real number set $\mathbb{R}$ perfectly: The horizontal straight line $\mathbb{L}$ is exactly covered by $\mathbb{R}$.

4 Complex Numbers

Before concluding this appendix of the real number system, let us ask the following question: Is the real number system $\mathbb{R}$ perfect? To answer this question, we look at the following equation:

$$x^2 + 1 = 0. \tag{4.1}$$

We could just say the equation does not have a root in $\mathbb{R}$. Therefore, we encounter an equation of second order that does not have (real) solutions. Recall the procedure of developing $\mathbb{R}$ from $\overline{\mathbb{N}}$ to $\mathbb{Z}$, then to $\mathbb{Q}$ and finally to $\mathbb{R}$, we could ask if it is possible to further extend $\mathbb{R}$. This can be done by introduce the root i of equation (4.1). Thus, i is a symbol that satisfies

$$i^2 = -1.$$

Then we introduce the following set:

$$\mathbb{C} = \big\{a + bi \mid a, b \in \mathbb{R}\big\}.$$

any element $a + bi \in \mathbb{C}$ is call a *complex number*, with a and bi called the *real part* and the *imaginary part*, respectively. Also, we define the *conjugate* of $(a + bi)$ by

$$\overline{a + bi} = a - bi,$$

and the *modulus*

$$|a + bi| = \sqrt{a^2 + b^2}.$$

On $\mathbb{C}$, we introduce the following operations, which are extensions of the corresponding ones from $\mathbb{R}$:

- Addition: $(a + bi) + (c + di) = (a + c) + (b + d)i$.

- Subtraction: $(a + bi) - (c + di) = (a - c) + (b - d)i$.

- Multiplication: $(a + bi) \cdot (c + di) = (ac - bd) + (bc + ad)i$.

- Division: $\dfrac{a + bi}{c + di} = \dfrac{(a + bi)(c - di)}{c^2 + d^2}$, if $c^2 + d^2 > 0$.

Further, it is possible to define exponentiation. We omit that here. To conclude, we state the following theorem called the *fundamental theorem of algebra* whose proof will not be presented here.

Theorem 4.1. *Every non-constant single-variable polynomial with complex coefficients has at least one complex root. Consequently, every such a polynomial of degree n has exactly n complex roots, counting multiplicity.*

From this result, we see that the set $\mathbb{C}$ is *algebraically closed*, which is perfect!

Bibliography

[1] J. M. Almira, *On Müntz type theorems, I. Surv. Apporx. Theory*, 3 (2007), 152–194.

[2] C. Arzelà, *Sulla integrazione per serie, Atti Acc. Lincei Rend.*, Rome, 1 (1885), 532–537, 596–599.

[3] R. S. Borden, *A Course in Advanced Calculus*, Dover, New York, 1998.

[4] D. S. Bridges, *Foundations of Real and Abstract Analysis*, Springer-Verlag, New York, 1998.

[5] J. Dieudonné, *Foundations of Modern Analysis*, Academic Press, New York, 1969.

[6] J. J. Duistermaat and J. A. C. Kolk, *Multidinmensional Real Analysis I: Differentiation; II: Integration*, (Translation from Dutch by J. P. van Braam Houckgeest), Cambridge Univ. Press, 2004.

[7] W. Fleming, *Functions of Several Variables*, 2nd Ed., Springer-Verlag, 1977.

[8] D. J. H. Garling, *A Course in Mathematical Analysis, Vol. I: Foundations and Elementary Real Analysis*, Cambridge Univ. Press, 2013.

[9] L. H. Loomis and S. Sternberg, *Advanced Calculus*, Jones and Bartlett, Boston, 1990.

[10] H. Lou, *Mathematical Analysis: Key Points, Difficulties and Extensions*, High Education Press, Beijing, China, 2020 (in Chinese).

[11] W. A. J. Luxemburg, *Arzela's dominated convergence theorem for Riemann integral*, Amer. Math. Monthly, 78 (1971), 970–979.

[12] W. Rudin, *Principle of Mathematical Analysis*, 3rd Ed., McGraw-Hill, 1976.

[13] B. S. W. Schröder, *Mathematical Analysis: A concise introduction*, Wiley-Interscience, 2008.

[14] T. Tao, *Analysis I,II*, 3rd Ed., Hindustan Book Agency, 2014.

[15]   A. E. Taylor and W. R. Mann, *Advanced Calculus, 3rd Edition,* John Wiley & Sons, New York, 1983.

[16]   J. K. Truss, *Foundations of Mathematical Analysis,* Clarendon Press-Oxford, 1997.

Index

.

Printed in the United States
by Baker & Taylor Publisher Services